The Open University

Block 4
Configuration and component design

Prepared on behalf of the Course Team by Chris Earl

T211 Design and designing

This publication forms part of an Open University course T211 *Design and designing*. Details of this and other Open University courses can be obtained from the Student Registration and Enquiry Service, The Open University, PO Box 197, Milton Keynes MK7 6BJ, United Kingdom: tel. +44 (0)845 300 60 90, email general-enquiries@open.ac.uk

Alternatively, you may visit the Open University website at http://www.open.ac.uk where you can learn more about the wide range of courses and packs offered at all levels by The Open University.

To purchase a selection of Open University course materials visit http://www.ouw.co.uk, or contact Open University Worldwide, Walton Hall, Milton Keynes MK7 6AA, United Kingdom for a brochure. tel. +44 (0)1908 858793; fax +44 (0)1908 858787; email ouw-customer-services@open.ac.uk

The Open University
Walton Hall, Milton Keynes
MK7 6AA

First published 2004. Second edition 2009.

Edited and designed by The Open University.

Printed and bound in the United Kingdom by the Charlesworth Group, Wakefield.

ISBN 978 0 7492 2080 8

3.1

Contents

Exercises

Introduction

How does a designer turn a concept into a well-defined configuration of components? This is what Block 4 explains.

This block looks at how concepts are translated into layouts for workable, usable and well-defined products. It is a complex phase of the design process. Not only do designers and design teams have to address difficult problems, they have to do this in ways that allow creative development of the concept. Practical constraints must be integrated with innovative ideas. It is a stage when testing ideas is as important as generating ideas.

Configuration design and component design are broken down into the four stages of:

1. identifying components;
2. configuring those components;
3. deciding how to connect those components – and maybe devising how components in a mechanism should move;
4. defining the shape, material, size and appearance of all parts of the product.

And at each stage the designer keeps in mind the environmental impact that the decisions may have.

Designers determine the choice of components and the way components are positioned in two dimensions or three dimensions. In addition to this the designer needs to isolate the functions asked for by the design brief and determine how they can be implemented. A question the designer keeps to the fore is, what properties of the components will deliver the required functions in the final layout? Among the properties that can be manipulated are material, size, shape, colour and surface finish.

To assist your progress you will continue with the *Modelling Workbook* and you will be introduced to some new resources in the form of the *Shapes and Space Workbook* and solidThinking, the course computer-aided design software.

Block 4 concludes with case studies that describe how practising designers have tackled design briefs in different sectors of industry. These studies help to flesh out the way designers convert a concept into a well-defined configuration of components.

Aims and learning outcomes

Aims

The aims of the five sections of the block are, successively:

1 to distinguish four stages between concept and layout – parts, arrangements, connections and form;

2 to explore the phenomenon of configuration design with reference to a variety of design contexts;

3 to define various types of function relevant to design and designing;

4 to explore the phenomenon of component design, exploring component properties including shape, size, strength, finish and colour;

5 to illustrate configuration design and component design in a variety of commercial design contexts.

Learning outcomes

Having studied this block, you should have achieved the following learning outcomes.

1 Knowledge and understanding

You should be aware of:

1.1 the four stages of progressing a design from concept to layout – parts, arrangements, connections and form;

1.2 configuration and components in typical products;

1.3 key types of function;

1.4 significant properties of components such as shape, size, strength, finish and colour;

1.5 basic sustainability issues and their importance in modern design practice.

2 Cognitive (thinking) skills

You should be able to:

2.1 identify parts, arrangements, and connections in basic products;

2.2 identify organising principles and rules useful in working towards a layout design;

2.3 define various types of function and distinguish between functions in the analysis of existing designs and the generation of new designs;

2.4 transfer knowledge of basic component properties from one application to another;

2.5 identify configuration design and component design in examples of professional design practice.

3 Key skills

You should be able to:

3.1 apply analytical skills in exploring selected examples of designed products;

3.2 apply knowledge from one design context to another;

3.3 apply communication skills in documenting and reporting findings;

3.4 apply skills in basic computer-aided design.

4 Practical skills

You should be able to:

4.1 apply drawing and other modelling skills in configuration design and component design.

defining the form. All these aspects of form are driven by the requirements for the design to perform in a particular way. Each component contributes to this performance.

Study Chart

Section	Main text	Workbooks	DVD	Block 4 Supplement
1	Concept to layout	*Modelling Workbook* Section 7	Workbook Activities 16–19 (video sequences); Mark Sanders (video sequence)	Study guide; Completing the *Modelling Workbook*; Using the DVD
2	Configurations	*Shapes and Space Workbook* Sections 1–5	James Dyson (video sequence); Repeating patterns (video sequence); Exploring configuration (software); Symmetry and everyday objects (software)	Starting the *Shapes and Space Workbook*
3	Function			
4	Component design	*Shapes and Space Workbook* Sections 6–8	solidThinking (software); creative product design (video sequence)	Using solidThinking to make simple CAD models
5	Four designers turn concepts into layouts			

Support notes for TMA 04 appear in *Block 4 Supplement*. You should refer to these as you prepare to answer TMA 04.

1 Concept to layout

This section opens with a closer look at the activities in the design process that create a layout from a concept. It will then identify four subsidiary stages and introduce some familiar products to illustrate activities in the four stages.

You have reached a halfway point in the course. So far you have seen examples of designs as well as ways of designing. With designs ranging from chairs to sewing machines, you now have a good understanding of why new designs are created. You learnt about the need for well-defined user requirements and the dissatisfaction a user experiences when requirements are not met.

concept
idea underpinning a design – usually represented by models, and the start point for configuration design

Being creative and having new ideas are hallmarks of designing. However, designing does not finish with a new concept. As you have seen in the first three blocks of this course, drawings or physical models are some of the ways to represent a concept. These are only the starting points in a process that leads to a new product that is manufactured, distributed and sold. The concept sets you off in the right direction, but for manufacture to be possible it is necessary to define the precise shape, size and arrangement of all the parts.

You know what a design concept is, but what is a layout? Concept design deals with ideas rather than precise forms. This block deals with the phase of designing that starts with a concept and finishes with a definite product form. To reach a 'definite product form' decisions have to be taken about how components will be arranged, how functions will be implemented, and how components will be shaped and coloured. The result is the definition of the design as a whole. I call the definition of the design as a whole a layout.

layout
result of configuration and component design – definition of components, arrangement of components, connections between components and their form

embodiment
another way to describe the layout – the embodiment of concept in the layout, and the result of configuration and component design

Another name for the phase I'm going to describe in this block is embodiment because a concept is embodied in a particular form. Because I just said in the paragraph above that the concept is converted into a layout, you can think of the concept being embodied in the layout.

Other features of components such as their colour and texture also contribute to defining the form. All these aspects of form are driven by the requirements for the design to perform in a particular way. Each component contributes to this performance.

A layout shows the spatial arrangement of all parts, including their shapes, sizes and other features. The layout, like a concept, may be made available as a drawing, a computer-aided design representation or a physical model.

There is a further design phase necessary to complete most design projects. This phase defines the 'deep' details of the components. Details include the precise specification of dimensions suitable for manufacturing, the materials from which they are made, the processes used in their manufacture, as well as sources of supply and channels of distribution. This next phase is called *detail design*. Block 4 will touch on some aspects of detail design but only because layout is a precursor to detail design.

1.1 Typical tasks between concept and product

There may be many satisfactory layouts of a concept that meet most user requirements. Likewise, there will be many that are not satisfactory. The layout is usually done in several stages with the design becoming more defined at each stage. This is a complex process in which the details of the design are progressively defined. To appreciate the whole range of things a designer does when creating a layout, it is helpful to consider the difference between a concept and the final design or product.

Exercise 1 Four strides

You have seen concept sketches for the Strida bicycle and you can see them again in Figure 1. Many possible concepts were explored based around the principle of folding into a stick-like object resembling an umbrella or baby buggy. The designer chose a basic triangular concept.

For this exercise I want you to suggest four major tasks the designer has completed between settling on the concept of the triangular frame and the final design of the production bicycle.

You may find it helpful to return to the video sequence on Mark Sanders and the Strida bicycle as you do this exercise.

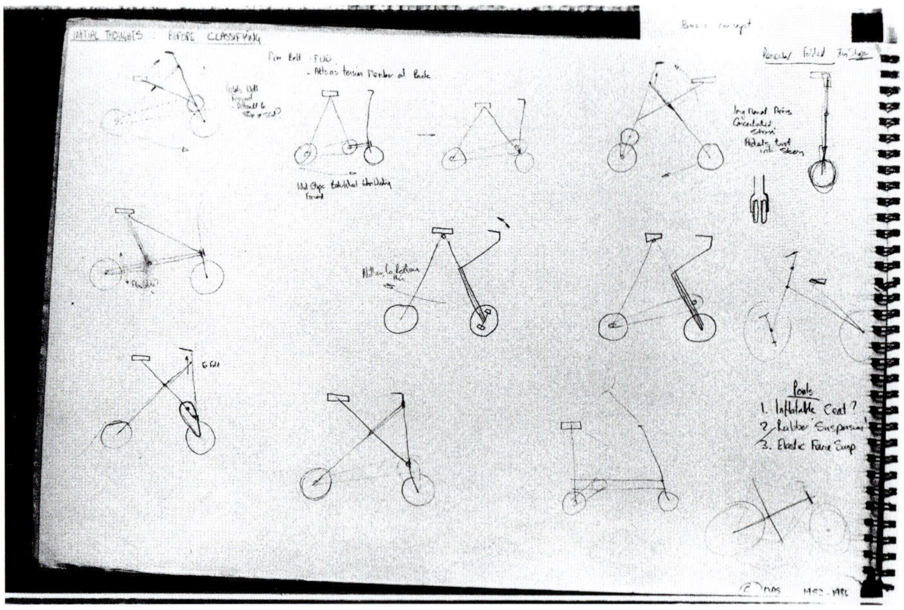

Figure 1 Sketches of Strida bicycle concept

Figure 2 Strida bicycle mark 2 and details of lower front removable connection, toothed belt and chainwheel

Discussion

The major parts of the bicycle were identified at the concept design stage. In most design projects there are many important design tasks and you may have picked different ones to me. For the Strida I pick out the following four tasks that were completed between concept and final design.

1 The first task was to decide how the frame was to fold up. The triangle needed to be disconnected. Mark Sanders decided to do this at the lower front (see Figure 2). Of course the frame then hinged at other points. There were subtasks in designing the hinges as well as the lower front connection. The latter had many of its own requirements – it was a design in its own right. It had to remain secure when the bike was being ridden; it had to be quickly released and reconnected; it had to resist damage from accidentally hitting the connecting pieces. The hinges needed to be specified.

2 It was decided to locate the pedals on the lower bar of the triangle driving the rear wheel. This meant the triangle could not be disconnected at the back wheel because the pedals and the wheel need to have a fixed length between them. A second task was to decide how the pedals would be connected to the rear wheel. In other bicycles a chain and toothed chainwheel are used. For the Strida a toothed belt replaced the chain. The designer had to ensure the belt would stay on the chainwheel by using guides.

3 A third task was to decide how many teeth should be on the chainwheel and how many on the smaller toothed gear attached to the rear wheel. These are important decisions because they determine how easy or hard the bicycle is to pedal.

4 A fourth task was to design the exact shapes and sizes of all parts. The shape and size of tube for the triangular frame needed to be decided. Decisions had to be taken about wheels, saddle, handlebars, pedals, and so on.

Of course there are other tasks. One essential task not mentioned above is the choice of material from which the frame of the Strida bicycle is made. The designer cannot leave this decision until the end of the design process. The exact shapes and sizes of all parts – the fourth task in the exercise – cannot be determined independently of choice of material. You will find this point reiterated at several places in this block, namely that shapes and sizes cannot be determined independently of material and manufacture (how to choose materials is looked at in Block 5). In any design project it is essential to consider all the properties of parts in a balanced way.

In doing Exercise 1 you saw that there are many activities to be completed between concept and the final design of a product. Many of these activities are complicated tasks in their own right.

Recall that in the Strida DVD video sequence, materials are discussed alongside methods and costs of manufacture in the design process.

1.2 Four stages in creating a layout

Suppose you have an idea for a new product and have made some sketches. Perhaps these show how the idea might work, how it might be used or how the product will be shaped and coloured. The sketches may also indicate the design has several parts. For example, a bicycle has a frame, pedals, wheels, and so on. A chair usually has legs, seat and back – although you should notice that some chairs, like the Verner Panton chair shown in Section 5 of Block 1, are made from just one part.

Exercise 2 Parts

With reference to household items in a kitchen, identify a product with just one part, a product with several (less than 20) parts and a third with many parts (up to 100).

Discussion

Figure 3 China cup, plastic cup, and spoon

You maybe thought of a plastic bowl or a ceramic saucer that usually has just one part. Although a china cup does have a handle and a body, they are fixed rigidly together and have been fused together in firing the clay. For plastic cups the whole cup and handle are made as one piece in a mould.

The spoon in Figure 3 is another product with a single part – again this has a handle and bowl but they are formed from one piece of metal. You see that identifying how many parts a product has is not always easy.

I could say that a part is formed by a single piece of material that has always been a single piece all through production. If this were the definition of a part, the spoon and the saucer are single parts while the china cup with a handle has two parts at the moulding stage of its manufacture, but only one part after firing.

An electric kettle has several parts. It may easily have 20 parts when the separate parts of the switches and plugs are added to the parts of the kettle body, heating element and lid. It is surprising how few products have between 2 and 20 parts, especially when you really count all the separate parts, however small. You might like to estimate the number of parts in your own kettle without including the plug. Start by listing the parts that you can see. Then make a guess as to parts you cannot see – do not take your kettle apart to do this. In addition, recall the example from Block 1 of the flat-pack chair from IKEA. There are several main parts but also some other parts – special fixings to join the chair together. Some of these are supplied in multiple numbers. That chair has over 20 components.

In the category of products with many parts, a washing machine, microwave oven or cassette-radio are examples you might have in your kitchen. In these there are complicated assemblies of switches, motors and electronics. The motor for rotating the washing machine drum or the pump for emptying the water are made from many parts. These subassemblies perform a specific function and are sometimes called *functional subassemblies*.

A sketch for a concept design may identify the main parts. Different concepts have different parts and different arrangements of the parts. For example, look at sketches for possible concepts of the Strida bicycle in Figure 1. These sketches show the main parts and possible arrangements.

However, in creating a layout the designer defines all the parts and their arrangements. As you saw in Exercise 2 products often have many parts. Defining the layout can be complex and being systematic helps.

In trying to be systematic I've divided the process of creating layout from a concept into four stages. These are not hard and fast stages in practice because it may be difficult to isolate the stages. I will use the four stages as a helpful guide.

1.2.1 Stage 1 – define main parts

component
item from which products are constructed

In the first stage the main *parts* (components) that were identified in the concept are defined more precisely. These are the main, functional parts that must be present if the concept is to be realised and user requirements satisfied. For example, in the Strida bicycle, the bars of the triangular frame, seat, handlebars, pedals and chainwheel were the main parts identified in the concept sketches.

The parts of the Strida were given approximate shapes and sizes. For example, the distance between saddle and handlebars, the size of tube to support loads when riding and gearing ratios are all estimated roughly.

For a knitted jumper the parts in the concept may be a patterned front, back, sleeves and neck but they might not appear separated in the overall concept sketch. More likely they will merge, one into another.

Also in the first stage the purpose of each of the main parts is defined. However, often the purpose is implicit and it will be obvious from the context. In the case of the Strida it is obvious what the saddle, pedals and handlebars are for. But it may be necessary to be more explicit about the purpose of some parts. For example, the front side of the triangle will include the steering.

1.2.2 Stage 2 – explore arrangements

With the main parts better defined, the second stage explores their possible arrangements. One possible arrangement of parts is called a configuration.

configuration
a representation that identifies the main parts of a design and their arrangement

A concept sketch may give some help here but now the designer searches for different ways to arrange the parts and realise the concept differently.

In the case of the knitted jumper there may not seem much scope for this – front is front and sleeves are sleeves. But even when arrangements of parts seem tightly constrained there are usually some opportunities for creativity, albeit at a finer level of detail. For example, different patterns of knitting can be explored for creating the effect specified in the concept. At this stage possible arrangements of parts are explored subject to constraints on the whole design. You saw in Block 2 that there are two kinds of constraint, namely the firm or 'hard' constraints that must be satisfied and those 'soft' constraints that are desirable but not essential.

1.2.3 Stage 3 – define connections

connection
interface between a component and other parts of a product

In the third stage the designer explores how the main parts are joined together. This is a stage concerned with defining connections. It may introduce new parts like screws, bolts, glue or clips; or it may specify movement between connected parts.

1.2.4 Stage 4 – decide on shape, material and appearance

form
the sizes, shapes, colour and surface finish of each component are determined to give a well-defined, whole product

In the fourth stage the form can be defined. That is, the sizes, shapes and possibly other properties of a product, like colour and surface finish of each part, are determined. There is still scope to explore possibilities, for example exploring different shapes.

By the end of the fourth stage a design proposal will have well-defined parts joined together in specified ways. Concept designs that may have been rough and impressionistic have become a layout. The design is more definite and might be presented in a number of ways, including accurate drawings and computer models.

1.2.5 Assess possibilities during stages

Each stage is an exploration of possibilities. For example, the third stage explores how the main parts can be connected and specifies any new parts that are needed to form the connections. In the example of the Strida the connection point on the frame above the front wheel requires a special pin and catch (Figure 2). These new parts may have several possible arrangements that create the desired connection between the parts. So the second stage is not the only stage at which arrangements of parts are explored.

At each stage the possibilities are assessed in terms of how they meet the specifications for the design. If there are many possibilities, those that perform best are chosen for further development. Modelling is used to evaluate the design proposals.

However, it may happen that none of the possibilities meets the specification. In this case the designer creates alternative possibilities – perhaps modifying a possibility that nearly meets the specification. To do this it may be necessary to go back to an earlier stage, perhaps to find an alternative arrangement of parts. This will involve iteration through the stages.

1.2.6 Configuration design and component design

configuration design
main parts are identified and arranged to produce a configuration

component design
process of defining the form of components and the connections between them

The picture you now have is four stages, with assessment against specification at each stage and iteration to improve the design. Together, stages 1 and 2, which identify the main parts and their arrangements, are called configuration design. Together, stages 3 and 4, which is where all the parts and connections are decided on, are called component design.

When all the parts have their shapes and sizes defined, the arrangement selected, the connections developed, and the overall form generated, a *layout* for the design is established. I've represented the stages as the diagram in Figure 4.

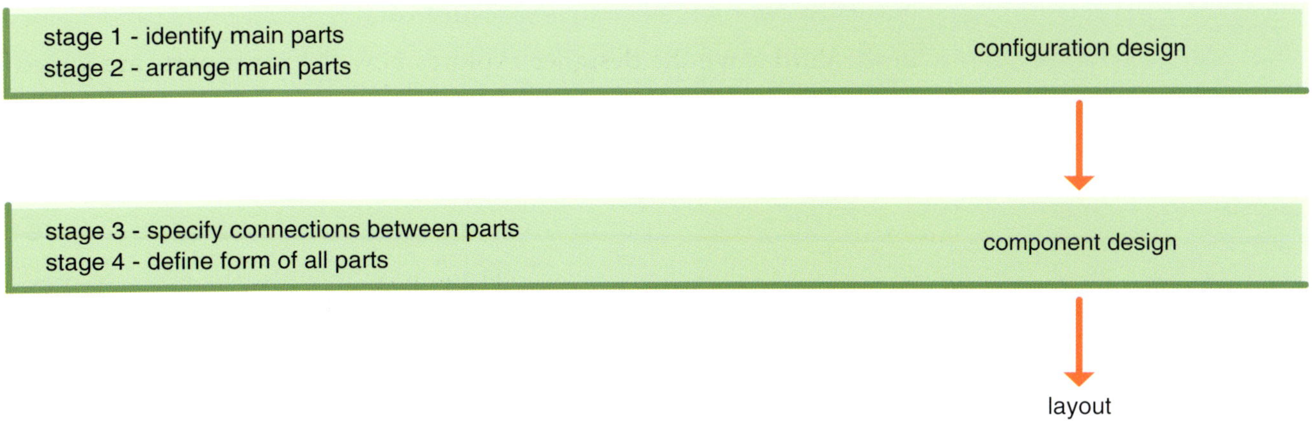

Figure 4 Four stages leading to a layout

Figure 5 Layout of portable electric drill

In configuration design designers specify which parts are to be connected to which others. In component design the tasks focus on the specification of sizes, shapes and the ways they are connected. Configuration and component design together result in a layout.

Although there may be many possible configurations that realise a concept at the end of stage 2, by the end of stage 4 these possibilities have been narrowed to one layout design. A designer will make many decisions in narrowing down possibilities. Some of these decisions are about how well a design will work for its users – how it will meet the user needs – while others are about appearance and cost, for example.

1.2.7 Layout puts things in their place

Layouts specify how all parts are located relative to one another. Therefore layouts for products are essentially three-dimensional. They are represented in many ways including physical models, two-dimensional sectional drawings, perspective drawings or as computer-aided design (CAD) models. Figure 5 shows a layout for a portable electric drill that reveals all the parts with specific and definite sizes. This is derived from a physical model for an electric drill and represents a cut-away section through the middle of the drill.

A layout for the electric drill is complicated. There are shafts, bearings, motor windings, switches, electronics, and electrical connections. The layout representation shows how all these parts are arranged and their shapes and sizes. Further, for some parts the details are not included in the layout. For example, bearings will be specified by a reference to a specialist catalogue. Layouts can also be represented by drawings showing sections, exploded views, CAD models and prototypes. I have used the cut-away as one way to represent a layout for the electric drill.

Exercise 3 Distinguishing concept from layout

From the first three blocks of the course, pick out an example of a concept sketch. List three differences between a concept sketch and the layout of the electric drill in Figure 5.

Discussion

A concept sketch gives an impression and expresses an idea. Examples in earlier blocks include the sketches of cutlery, lawnmowers or cars. The concept sketch picks out a few aspects of the design that are important for the user or customer. In the case of the car these might be styling features.

The layout of the electric drill shows details of all the parts and shows how the design actually works, though admittedly it might be hard to read this directly from the layout. I've picked the following three differences.

1. In the layout *all* parts are shown whereas in the concept sketch only some of the parts might be drawn.
2. In the concept sketch parts are shown approximately in shape and size but in the layout shapes are represented accurately and to scale.
3. From the layout it is possible to understand precisely how the design will work. For example, the specifications of motor, gearing and electronic controls define how much torque the drill will deliver at different speeds.

1.2.8 Layout leads to the detail design phase

A product is not manufactured directly from the layout. The layout represents the configuration of parts, their shapes and their sizes. However, manufacture requires the specification of all properties of all parts – which is done in detail design, the phase after layout.

Detailed designs have dimensions given in a special way that specifies how accurately a part needs to be manufactured, that is, its tolerances. You will read more about tolerances and detail design in Block 5. The detailed designs for components also specify the materials from which the parts are made and their methods of manufacture. They may be computer representations that can be passed directly to the computer-controlled machines that make the parts.

The distinctions between creating a layout design and a detail design are not always clear in practice. Take the example of deciding on materials for components for the electric drill. Without a specification of the type of material to be used for each component it would be impossible to decide on shape, size, and methods of joining. Although final decisions on materials may not be made until after the layout is defined, a designer needs a good idea of the types of material and their properties – for example strength, flexibility and hardness – that are going to be used. This is especially true where a new material can, for example, allow new shapes or provide a more suitable surface texture for a component.

So it is not possible to leave all decisions on manufacture and assembly until after the layout is completed. The arrangement of parts in a layout will be poor if it is not physically possible to assemble the parts. Furthermore the designer at the layout stage must think even further ahead to disassembly, that is, how the product can be taken apart. Why is this so important? One reason is for maintenance. Another, as you saw in earlier blocks, is new regulations on the recycling and reuse of parts and materials of products once they have reached the end of their useful life.

If the parts are to be recycled, it is essential the parts made from different materials can be separated, perhaps by disassembly. In this way materials can be recycled in relatively pure form. If parts are difficult to disassemble or contain mixtures of many materials, any recycled or recovered material will be contaminated. Recycling is just one part of the wider issue of sustainability of a product. You have already encountered several sustainability issues in the course.

The stages in creating a layout amount to one phase in the whole process of product development from analysing user needs and writing a design brief, through concept, layout and detail design to manufactured product. This process of product development involves transformations, taking user needs and transforming them through a sequence of phases into a real product that people can use.

SAQ 1

Identify three pieces of information that a layout contains.

1.3 Developing a layout

In the worked example to TMA 01 in the *Block 1 Supplement*, you saw the concept design of a small domestic table. I'm going to return to this context to develop this section.

1.3.1 Going from concept to table

Suppose you are designing a coffee table to put your drinks on or to leave your newspapers on, with a shelf below perhaps for older newspapers. First you might sketch an idea that gives the feel of the style, shape and general effect.

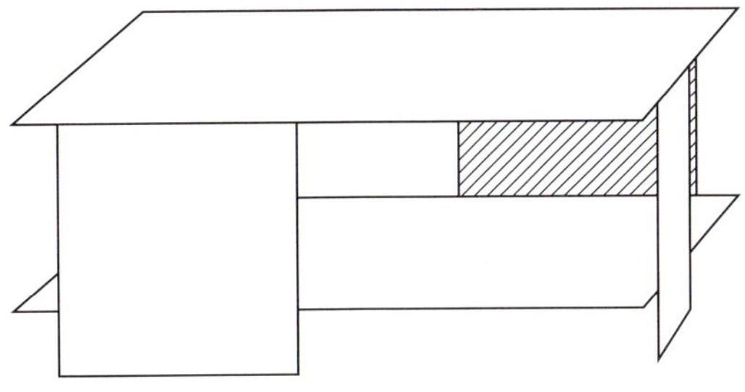

Figure 6 Concept sketch of table

Figure 6 shows a sketch I made during the process of creating concept designs for the table. Figure 7 shows pieces of furniture designed by architect Rudolph Schindler in the 1920s and 1930s for houses he designed in Los Angeles. These historical examples helped in forming my concept sketch. As you saw in Block 3 there may be many sources of inspiration. Designers often borrow, modify or work in a particular style. I don't feel guilty about borrowing ideas, and neither should you.

Figure 7 Furniture designed by Rudolf Schindler

1.3.2 Table parts

This particular table concept provides a good subject to explore layout design, even though it's a different concept to the one selected in the supplement to Block 1. The first stage of the four described above is to identify the main parts. For this concept of a coffee table there are six main parts: top, four legs and a shelf. At this stage it is desired that each part can be cut from a single sheet of plywood in the style of Schindler's furniture. This may not be possible in the end but it's on my wish list. Even at this early stage it is possible to assess the concept. The plywood will make good surfaces for the top and shelf and it seems likely the four legs will easily support the objects likely to be on the top and shelf.

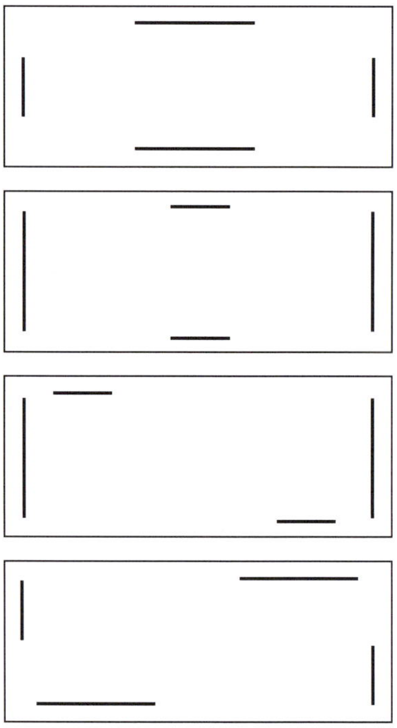

Figure 8 Four possible arrangements of legs on the coffee table

1.3.3 Arranging table parts

Stage two is arranging the main parts. Consider arranging the four legs. There are many possibilities within the concept. The legs could be arranged underneath the tabletop in any of the ways in Figure 8 – all slightly different. Which arrangement would you choose?

You need a criterion to assess a good arrangement. One criterion might be that you should be able to see what is on the shelf and get access to it easily. Another criterion is stability. Having legs in the middle of the long sides would not be effective against either criterion. In the end I chose one possibility: legs towards the corners (see Figure 8, bottom arrangement).

1.3.4 Connecting table parts

Now, in stage three, I will specify a way to join the parts together, that is, their connections.

Exercise 4 Join me

This is an exercise about joining parts together. For my design of a coffee table I decide to make it all from pieces cut from a sheet of plywood. In this exercise I could use any reasonable thickness of plywood – I will specify a sheet thickness of 18 mm. The top, shelf and legs are all rectangular pieces of plywood. I am going to get the pieces of plywood machine cut, so they will have square edges. Suggest how I can join the pieces together.

Discussion

Notice first that the material has been specified as plywood. Furthermore the parts of the table are going to be made by being machine sawn from a large sheet. These are issues associated with detail design rather than layout design. This overlapping of detail and layout design indicates these phases cannot be separated neatly in practice.

The overall design ensures the strength of the table does not rely just on the strength of the joints between legs and tabletop. The joints between legs and the shelf pull the table together and make it rigid. Probably glue by itself would not be sufficient for holding the parts together. So that the tabletop surface is not marred by screws or nails, one possible option is to use small round pegs of wood (dowels) to strengthen the joint (see Figure 9).

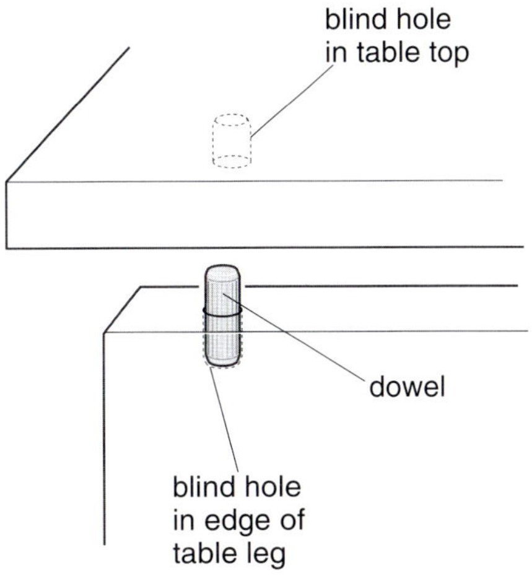

Figure 9 Fitting legs to the coffee table top with dowels

The dowels are inserted into shallow holes drilled in the edge of one sheet and the surface of another. These holes are referred to as blind because they do not go right through the material. This means that once assembled the dowels are entirely hidden inside the joints.

I have begun the processes of stage three – deciding how to connect parts together – but I should also assess whether the proposal to use dowels in the joints will work. Notice the dowels around the shelf are horizontal. The dowels in the joints between top and legs are vertical. These different directions help to make the table rigid. Further, with at least two dowels along each joint then possible movements at joints are minimised. I think this connection method will work.

1.3.5 Deciding on table shape, size, material and appearance

The fourth stage, to determine the product form leading to a layout, involves many decisions. Exactly how high, how long and how wide do I want the table to be? How big – diameter and length – are the dowels? Where should the dowels be placed in the joint? How thick is the sheet? This should be thick enough to give a rigid surface and allow the dowels to be fixed within the thickness. Plywood is made in sheets of standard thickness (from 3, 6, 9, 12, 15, 18 and 24 mm) and to a standard sheet size 2240 × 1120 mm. I have already proposed 18 mm plywood, so circular dowels that are 6 mm across and 24 mm long would seem appropriate. This allows space for the dowels in the joint, and the 18 mm plywood sheet will produce a strong coffee table. Each joint will have two dowels. Dimensions of the table – length, width and height – come from user needs for function (room space available, table area and suitable height) and appearance (proportions). Now I need to decide on the exact shapes and sizes of the pieces of plywood (see Figure 10 for the dimensions of the top and legs).

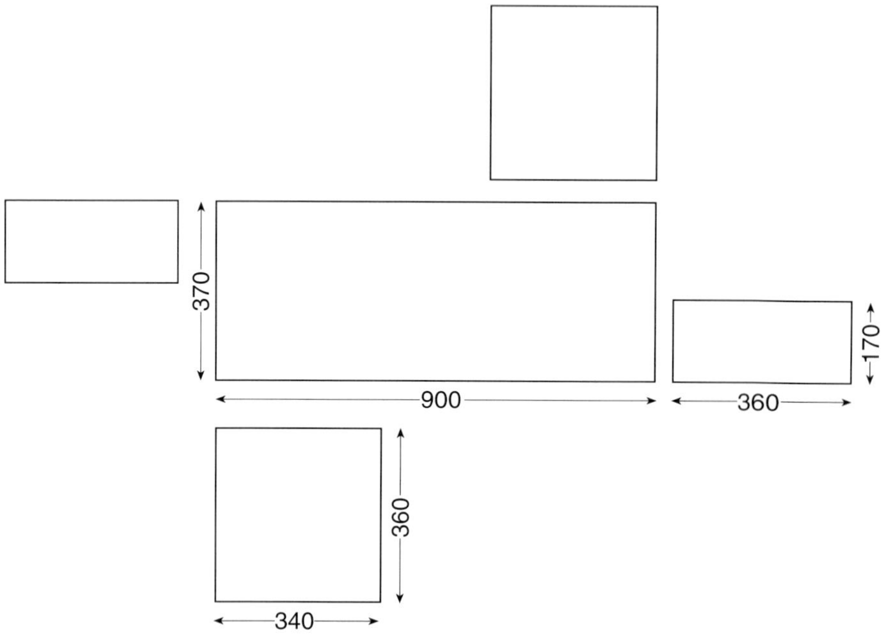

Figure 10 Parts of the table with dimensions in millimetres

I also need to think about what the table will look like. What kind of table surface will be used? The bare plywood could be varnished and polished.

Sections and views of the coffee table (Figure 11) show all the parts, and their arrangement. These particular drawings do not show how the parts are joined together. However, they clearly show the shapes and sizes of all parts.

How could I make the table? I have already decided to cut the parts from a single sheet. Should I make all these bits or get someone else to make them? Should I make the dowels or buy them ready made? Before I can begin making the table I need to know where to cut the sheet, where to drill holes, and the size of the holes. I need to cut out the parts for the table from a sheet of plywood without too much waste.

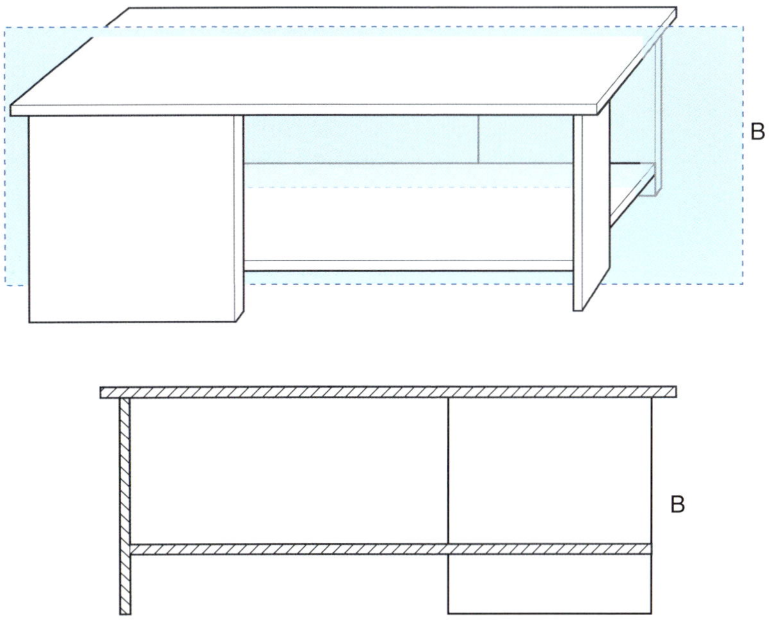

(a) Drawing of the final design with plane B and the corresponding section

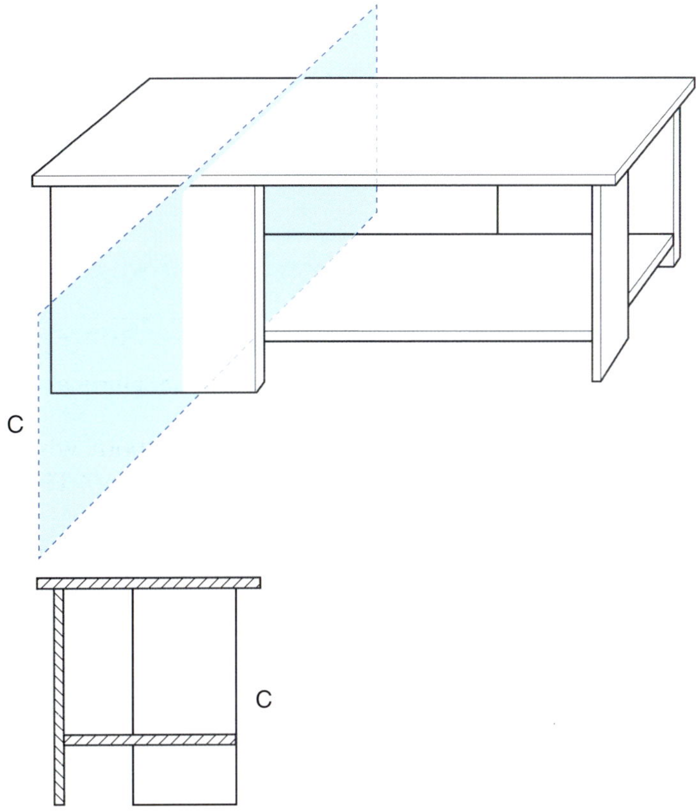

(b) Drawing of the final design with plane C and the corresponding section

(c) Horizontal section of table

Figure 11

Exercise 5 Another plywood table

Produce a perspective sketch of a plywood coffee table that is similar in concept to the coffee table above but with a *circular* top. It is to have four legs and an *octagonal* (eight sided) shelf cut out from a plywood sheet.

This exercise will give you practice in sketching a circular shape in perspective as an oval.

Discussion

The main parts of the table are still top, legs and shelf. The configuration and connections of parts I developed for the rectangular coffee table can be transferred to the circular table. One possibility is for the four legs to be connected to the octagonal shelf along alternate sides.

If you had trouble with the sketching in Exercise 5, I suggest you refer to Section 7 in the *Modelling Workbook*, which shows you how to construct perspective views of cylindrical shapes. The table top is a very short cylinder. Being able to sketch simple shapes such as cylinders in perspective is a useful skill for representing designs. If the sections constructed in Figure 11 above are still a mystery, I suggest that you take another look at Sections 4 and 5 in the *Modelling Workbook* now.

1.3.6 Representing parts and connections

Exercise 6 Making connections

A fashion sketch of a dress reveals the overall design and configuration of parts (Figure 12). List five different parts of the dress, such as the collar. For each part list the other parts it is attached to (configuration). How might you communicate these connections?

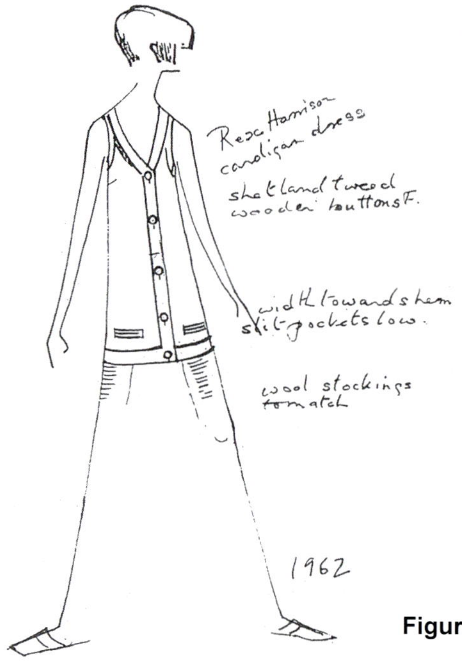

Figure 12 Dress made from several parts

Discussion

The five different parts could be collar, shoulders, body, buttons and pockets. The collar is attached to the shoulders, the buttons and pockets to the body and so on. You could represent the connections by drawing on top of Figure 12. Figure 13 is a diagram to show how I represented the connections. I put a dot on each part. I joined two dots by a line if the corresponding parts are attached to one another. Notice the positions of the dots in Figure 13 do not match the parts exactly but they are close enough.

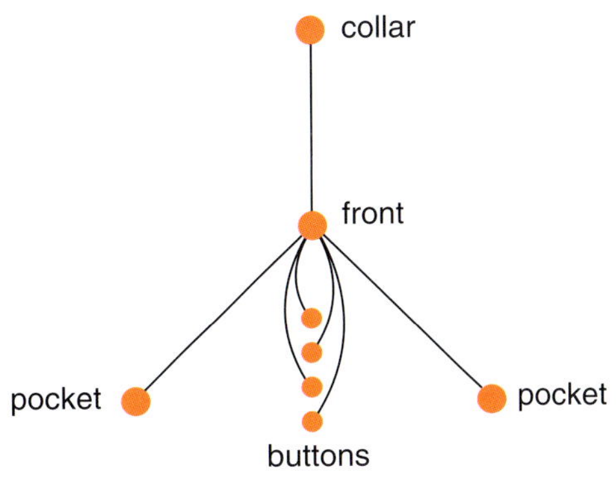

Figure 13 Representation of connections between parts of a dress

1.3.7 Information associated with a layout

But is configuration the only information that a layout provides? Recall from Blocks 2 and 3 that design concepts are based on principles. For example, a concept for a new bicycle exploits principles of mechanics and principles of gearing. The layout will show how the principles work in an actual physical device.

The layout design should contain enough information to allow prediction of how the design will perform and to what extent it will meet user requirements. For the coffee table example above the top surface is for drinks and the shelf for newspapers. From the layout you should be able to predict whether it will, for example, be rigid enough not to wobble and spill drinks.

You probably noticed that the information about a design differs during the concept and layout phases. Information on user needs, principles and function and a rough idea of appearance are included at concept phase. At the layout phase there is detailed information on parts, configuration, geometry, connection, working principles and performance. A way to think of this is that the concept phase provides a rough starting point for layout. The layout, in turn, is suitable as a starting point for the detail design phase.

You have seen the four stages in creating a layout design for a simple coffee table. This begins with a concept sketch and ends with a layout drawing. In this example evaluating the design against user requirements and product specification was straightforward at each stage.

1.4 Evaluating a familiar layout

I will now look more closely at how a design is evaluated to ensure it meets user requirements. To do this I'm going to examine arrangements of furniture in a room to show the importance of evaluation in creating a satisfactory layout.

Because a layout contains information on parts, arrangements, connections and form it can be evaluated to see whether it meets requirements of the specification. If the layout does not meet requirements, it can be changed and evaluated again before committing time and money to making the product. This process of iteration may occur several times before a satisfactory layout is finalised.

Think of a room where you sit to watch the TV or read the paper. In Figure 14 a layout of one such room is drawn in plan view with TV, bookcase, sofa, two chairs and coffee table. A perspective view of the layout is shown alongside the plan view.

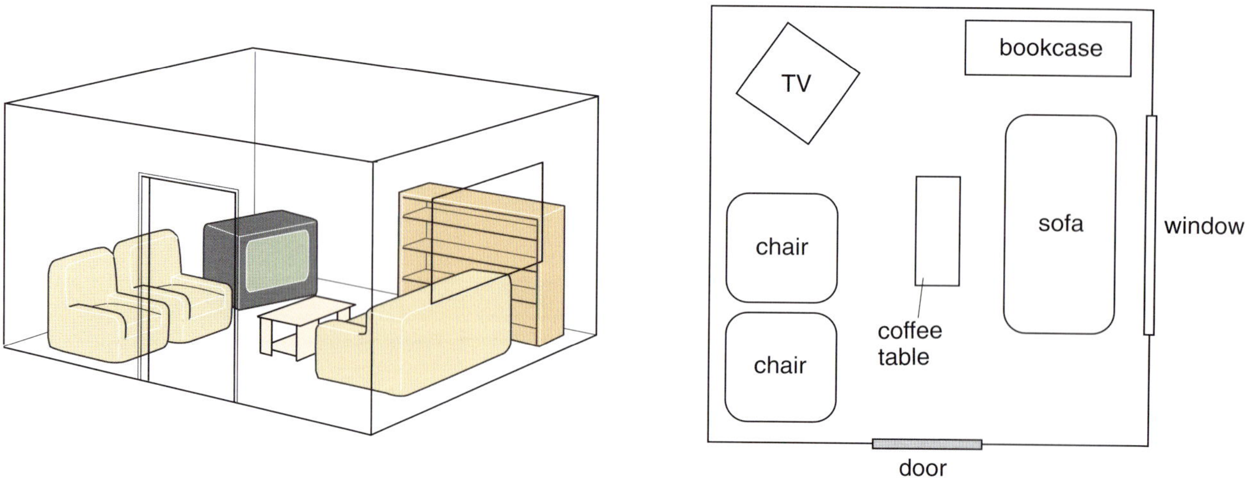

Figure 14 Room in perspective and plan views

All the items have specific dimensions (with height the third dimension) as do the floor area, window and door. The room is 4×4 m and 2.5 m high; the window is 2×1 m in the centre of the wall and starts 1 m above the floor; the door is 2 m high and 1 m wide. The layout shows positions of all items. Of course objects will move about slightly over time, perhaps chairs or the TV, but the configuration remains the same. The sizes of items are given in terms of a box-like envelope in which they fit (a bit like the imaginary crates you defined to help you draw objects in Section 6 of the *Modelling Workbook*). The chair is $1 \times 1 \times 1$ m, the sofa is $1 \times 2 \times 1$ m, the TV is $800 \times 800 \times 1000$ mm, and the bookcase $1500 \times 500 \times 1500$ mm (in each case the third dimension is the height).

Suppose that you want to reorganise the furniture in the room. In effect you are designing a new layout. What do you want to improve about the layout? Suppose the window looks out onto a garden and you want to increase the number of seats in the room that have a direct view of the garden through the window. This is going to be the basis for an evaluation.

Constraints apply to any changes to the layout. These constraints include:

- each item of furniture will have a free space around it for access;
- occupants of all chairs are to be easily visible to each other, chairs either have direct line of sight to the TV or a direct view out of the window, or both;
- people need enough space to move around the room especially to go in and out.

I show a drawing of the room that has rectangles containing crosses in front of each object indicating access space (Figure 15). Lines of sight from seats to TV are shown as green lines and lines of sight from seats to window by blue lines. Finally the paths from the door to each of the four seats are red curves.

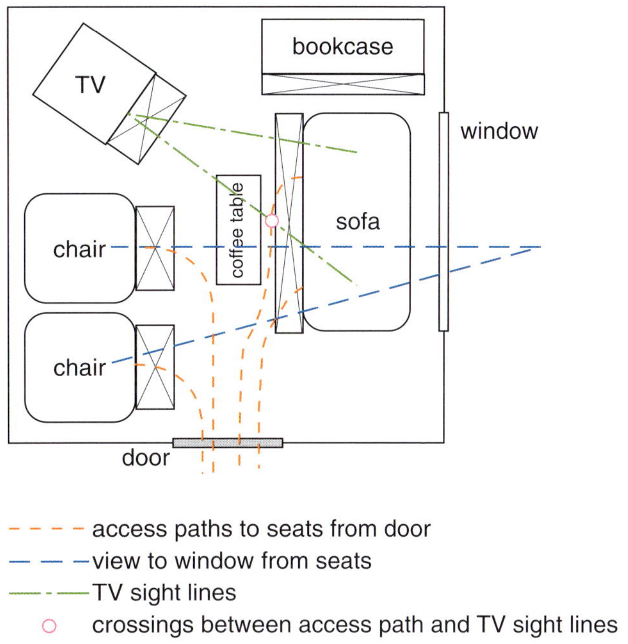

Figure 15 Room layout with access paths, lines of sight and crossing points

One way to get a more satisfactory layout is to experiment by moving the furniture around the room. A more designerly way is to draw possible layouts and evaluate them. Computer-aided design software can help as can other models, perhaps cardboard cut-outs of the outlines of the furniture on squared paper. The key point is that you generate possible layouts. Then you *evaluate* them against user requirements. Moving the actual furniture into different positions is not necessary.

Evaluation will keep to the limited assessment of access paths to the chairs and lines of sight. Evaluation will be based on the following calculations.

(a) The number of crossings between paths and lines of sight to the TV.

(b) The number of seats with a direct view of the window – this is the critical requirement.

(c) The number of seats with a direct view of the TV.

(d) The number of seats from which the coffee table can be reached easily.

Notice that (a) is a negative measure with more favourable designs having a lower value, while (b) and (c) are positive with more favourable designs having a higher value. This kind of evaluation does not take into account appearance and the feel of the room.

The existing layout has evaluation against the four measures (a), (b), (c) and (d) of (1, 2, 2, 3), that is, (a) one crossing, (b) two chairs with direct window view, (c) two chairs with direct TV view and (d) three chair seats are next to the coffee table.

Exercise 7 A better view

Try and rearrange the furniture to give more chairs with direct view of the window and TV but perhaps allowing more crossings of lines of sight.

Discussion

It would be inappropriate for the paths going in and out of the room to cross the line of sight between sofa and TV – so TVs are often placed on the other side of the room from the door.

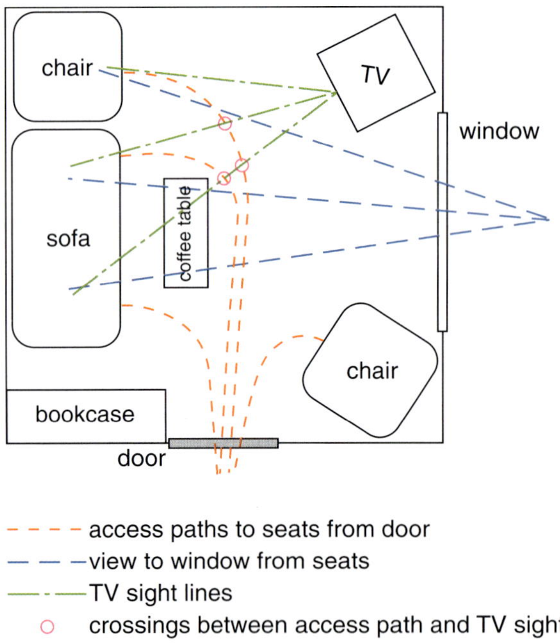

Figure 16 Room with alternative access paths, lines of sight and crossing points

It is often difficult to improve all measures at the same time – some trade-off may be needed. Figure 16 shows a possible arrangement with evaluation (3, 3, 3, 2). With reference to my list of (a) to (d) above, there are now three crossings (a), three chairs have direct views through the window (b), three chairs have direct views of the TV (c), and only two chairs are next to the coffee table (d). This is not an improvement against all the measures, but it does improve the critical measure (for this layout) of direct views of the window.

In Exercise 7, where a new layout was generated and evaluated, the layout was not generated systematically, it was created intuitively from a feeling of what might fulfil the specification. However, in creating the new layout I have, perhaps without being aware, limited the possible layouts by keeping to constraints of where items of furniture will be placed in the room and how they are located relative to one another. For example, I might want all the chairs to face one another, the bookcase against a wall and the coffee table in the centre of the room.

The four stages of creating a layout from an initial concept do not necessarily proceed as sequentially as my four-stage model might suggest. If evaluation of a proposed layout indicates an unsatisfactory design, the first thing to do is to change shapes and sizes, but it may be necessary to go back to look for alternative configurations. The whole process is iterative. This iteration requires both evaluation of a proposal and generation of new proposals and can take place at any stage.

1.5 Possible layouts

One of the designer's tasks is to explore possible layouts within the constraints to find one that meets user requirements. In Block 3 you looked at a number of techniques for assisting creativity, and you can apply what you learnt to generate possibilities for room layouts.

1.5.1 Necklace rule for room layout

One thing you might notice in the room layout exercise is that for this room full of furniture most is arranged around the walls – with the coffee table in the centre. One way of exploring possibilities is to look at sequences of items placed around the room like beads on a necklace. You could generate various sequences for the furniture as if you were arranging different types of beads in a necklace. Of course you would still need to evaluate each arrangement to see whether the furniture could be fitted in around the room. This necklace rule is a guide for exploring possibilities, but it does limit them. For larger rooms the necklace rule probably wouldn't help in generating possibilities as furniture can also be arranged in clusters in different parts of the room.

SAQ 2

What are the two essential activities for an iterative process like design?

1.5.2 Specimen designs

Another way to generate possibilities is to have several specimen layouts that are modified to fit the particular room and furniture. For smaller rooms you will probably have experienced the way that constraints limit you to a small number of possible layouts. A good example of this is a bathroom, which is often small with standard size items such as basin, bath, toilet and shower (Figure 17). For design situations like this it can be relatively straightforward to generate all the realistic arrangements of items because of the constraints.

For a small bathroom the constraints include ergonomic practicalities, safety regulations and restrictions imposed by the location of water pipes and drainage. Once you have generated the specimen layouts you can then choose the one that best fits your particular constraints and preferences and adapt it if necessary.

As the size of the bathroom increases and the number of items increases – bidet, separate bath and shower, and so on – possible layouts will change. Generally there are more possibilities as more space is available.

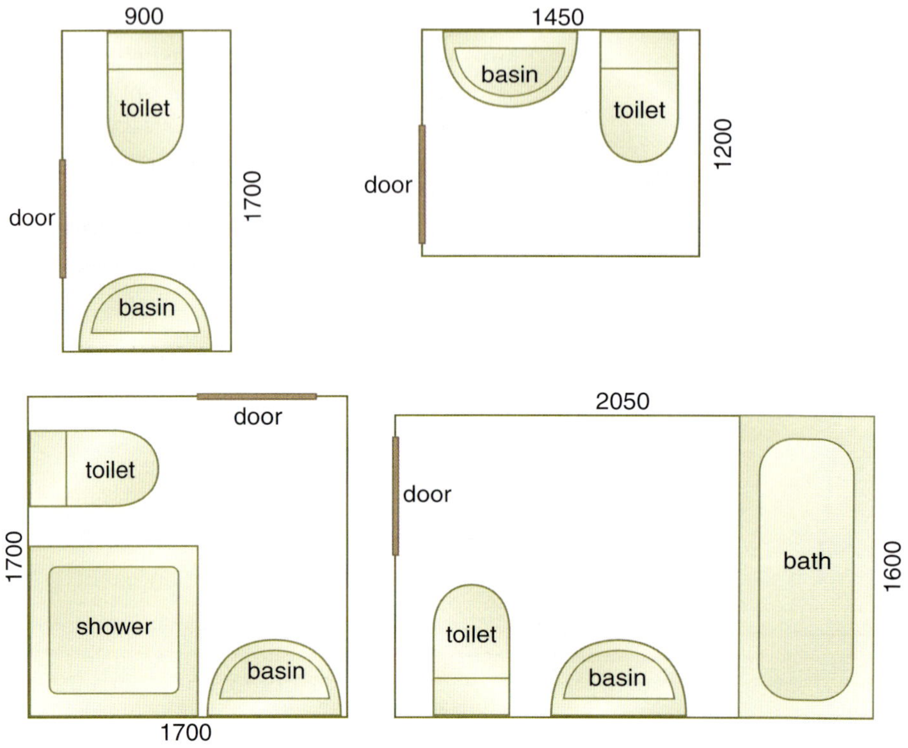

Figure 17 Specimen layouts for bathrooms (dimensions in millimetres)

The specimen bathroom layouts in Figure 17 summarise professional knowledge and experience. They do not contain extensive explanation of why the various possibilities are included. Nor are particular layouts identified as better than others. Specimen layouts show possibilities explicitly. The specimen layouts are the result of going through all the four stages of layout design – identify parts, arrange parts, join parts to the water supply and drains, and specify shapes and sizes in the overall form.

1.5.3 Standards

Published standards provide another type of guide for designers. They can help during all phases and are particularly useful when developing the layout. Standards are based on extensive experience and professional knowledge. There are national and international standards. The UK, Germany and the USA have national standards organisations: respectively the BSI (British Standards Institution), the Deutsches Institut für Normung (whose standards are called DIN) and the American National Standards Institute (ANSI). International standards are defined by the International Organization for Standardization (ISO). Each standard is referred to by a number prefixed by the organisation that issues it. To give an example, BS 2037 (1994) is a BSI standard for the design of portable aluminium ladders.

Figure 18 is just the contents page from BS 2037 (1994) *Specification for portable aluminium ladders, steps, trestles and lightweight stagings*. The parts, arrangements and properties are prescribed in the main sections of the standard, whereas the procedures for testing are given in annexes. Section 1 defines types of ladder in terms of arrangement of parts and conditions of use. Section 2 includes the properties that different types of ladder need to possess, and refers to the annexes that explain how to test that a ladder design has the required properties.

How do the design standards help in each of the stages of layout design for the ladder? For stage one of ladder design the main parts are identified – side pieces, steps, platforms, support rails (see Figure 18). For the second stage standard arrangements of these parts are specified for different ladders – step ladder, fixed-length ladder and extending ladder. The standard also helps at stages 3 and 4 by prescribing the properties the ladder should have. It lays down exactly how the ladder has to perform and gives a precise procedure for evaluating a layout.

BS 2037 : 1994

Contents

Figure 18 Contents page from the BS 2037 ladder standard

Standards express professional knowledge in a way that is easy to apply to a new design. Applying the standard guarantees the performance of the product at least in the properties the standard covers. Many products are not covered by standards. However, most products will have parts that are covered by a standard. For example, electrical components must meet safety standards for wiring and insulation.

1.5.4 Making use of rules, specimens and standards

In this section you have covered three different ways of helping to create possible layouts.

- First, there are rules to generate possibilities. One example is described as the necklace rule for furniture in a small room. Generating possibilities by rules will be examined in the next section.
- Second, for some types of design such as the fittings in a bathroom, there are specimen layouts.
- Third, guidance for creating layouts is provided in design standards. These give procedures to follow and tests to conduct that guarantee the product will be safe to use and meets basic requirements common to all users.

Exercise 8 Standard choices

(a) Look at Figure 19, which is standard BS 3692 (2001) *ISO metric precision hexagon bolts, screws and nuts*. The extract shown is the contents page. What do you notice is different about this standard as compared with BS 2037 for ladders?

(b) What types of product or design are likely to have official design standards (for example published by BS, DIN or ANSI)?

Discussion

(a) The standard for bolts prescribes the precise sizes and shapes for a selected range of nuts and bolts. The performance of each standard bolt is specified and procedures to test performance are given in annexes. The standard gives a limited range from which a designer chooses, but there are sufficient possibilities to cover most needs.

The ladder standard does not prescribe such a set of 'standard' designs but concentrates on the performance a design should have. Designs within the limits on size and shape laid down in the standard that have required performance meet the standard.

(b) Design standards are often used when safety is important, as with ladders and bolts. In the cases where each set of circumstances needs a special design then design standards give procedures to follow. In other cases, such as bolts, design standards provide a restricted set of possibilities and prescribe that only these exact possibilities are used. Designs, such as those for bolts, are used repeatedly in many different products and the standard enforces commonality, simplifies choice and ensures safety.

BS 3692:2001

Contents

		Page
Committees responsible		Inside front cover
Foreword		iii
1	Scope	1
2	Normative references	1
3	Finishes	2
4	General dimensions	2
5	Length of bolts and screws	2
6	Ends of bolts and screws	2
7	Screw threads	2
8	Length of thread on bolts and screws	3
9	Angularity and eccentricity of bolts, screws and nuts	3
10	Chamfering, washer facing and countersinking	4
11	Strength grade designation system for steel bolts and screws	4
12	Material and manufacture of steel bolts and screws	4
13	Mechanical properties of steel bolts and screws	5
14	Strength grade designation system for steel nuts	5
15	Material and manufacture of steel nuts	6
16	Mechanical properties of steel nuts (excluding thin nuts)	6
17	Drilled bolts with split pin holes	6
18	Marking and identification (see also Annex E)	6
19	Inspection and testing	6
20	Complete designation for the purpose of an enquiry or order	7
Annex A (normative) Recommended gauge for checking squareness of thread to face nut (see clause 9)		18
Annex B (normative) Chemical composition of steel nuts (see 15.2)		18
Annex C (normative) Testing of mechanical properties of steel nuts (see clauses 16 and 19)		19
Annex D (informative) Association of bolt and screw lengths and diameters (see clause 8)		21
Annex E (informative) Basis for the derivation of tolerances		24
Annex F (informative) Manufacturers' recommended range of sizes		30

Figure 19 Contents page from BS 3692 for nuts and bolts

It is interesting to compare design standards with the specimen layouts. The latter will be guides to design, rather than the rules published in standards. Specimen designs can be modified to suit particular circumstances and are often the result of a long history over which many possibilities have been tried and tested.

1.6 Product layout

This section examines the possible layouts of parts for the design of a different product, namely, the front of a microwave oven. The layout of the parts such as door, controls and switches is not highly constrained. There are no lists of specimen designs or design standards on this kind of layout. The parts have many possible arrangements, shapes and sizes, and there are many possible layouts. For most manufacturers a particular microwave oven front should have an appearance that differentiates it from competing products, is easy to use, and will look attractive in a prominent position in the kitchen.

1.6.1 Possible layouts of microwave oven fronts

What are the parts to be configured in the layout of the front of a microwave oven? First there is a control 'panel'. Second a display usually placed inside the control panel. Third there is the door. The fourth and fifth parts are the handle and the glass window in the door. Some ovens do not have a handle – in these cases the door may be opened by pressing a button. In Figure 20 I label these parts (a) for the door, (b) for the control panel, (c) for the display, (d) for the handle, (e) for the window. I will not count individual buttons and knobs as separate parts because I'm dealing with the main parts on the front. Figure 20 shows one configuration of these parts. The control panel is to the right of the door, and the door hinges on the left, opening right to left. The handle is on the right-hand side of the door.

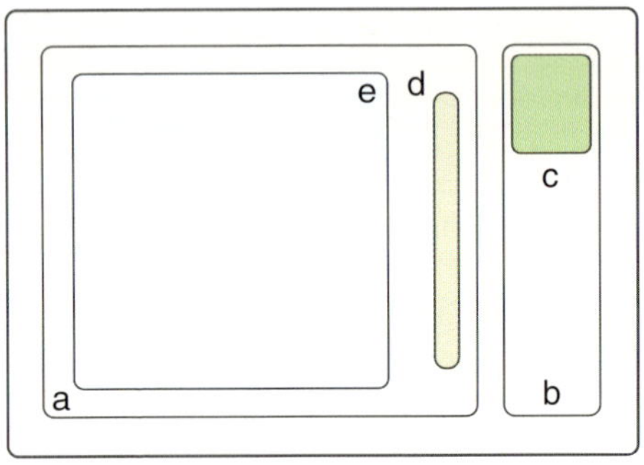

Figure 20 Configuration for the front of a microwave oven

The sketch of the configuration in Figure 20 is not how the product will appear – it does not contain details of shape, size, colour or surface texture. The coloured areas are just my way of picking out the position of the different parts. Figure 20 does show where the different parts are located in relation to each other.

With different materials such as stainless steel and coloured plastics and different shapes for the parts such as the glass window – microwaves can look quite different while retaining the same configuration. Of course there are also differences in the functions of different ovens. Some also have grills and convection ovens and some have weight sensors or even stirrers.

Figure 21 shows another configuration used for microwave ovens in fitted kitchens. Here the handle is at the top of the door, the door is hinged at the bottom, opening top to bottom. The control panel and display are located above the door along the top of the oven.

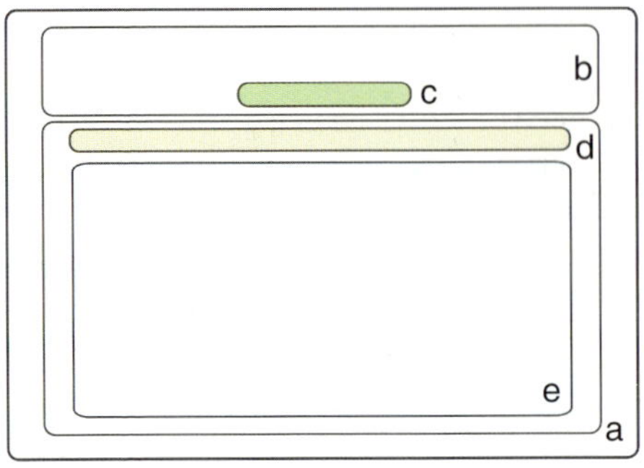

Figure 21 Another configuration for the front of a microwave oven

For each of these configurations there are many possibilities for different sizes and shapes. There could be variations in the outline shapes of the glass panel, or the small display panel, or the three-dimensional shapes of the handle. There could also be variations in colour and material. A final layout of the front of a microwave will include all these details and the connections between parts – showing how the window is set into the door or how the handle is fixed to the door, for example. Notice the four stages of layout design – identifying the main parts, arranging them, connecting the parts and finally specifying shape, size, colour and finish.

1.6.2 Evaluating product layouts

How are these layouts evaluated? First, appearance is important as microwaves are prominent items in the kitchen. Shape, proportion, surface finish and colour all influence appearance. Second, setting functions and timings should be straightforward. Third, opening, loading and closing the oven should be easy. Fourth, hygiene is critical in the kitchen, so shapes with surfaces that are easy to wipe clean should be used. This is one reason why press-button catches are used instead of separate handles.

Exercise 9 Configuration in microwave ovens

(a) Propose two more configurations for a microwave oven that are different from those in Figures 20 and 21.

(b) For one of the configurations draw a possible layout with this configuration. Your sketching for this exercise can be quite rough. Try and make your design interesting and different from those you see in the shops.

Discussion

(a) Notice that configurations for the two ovens in Figures 20 and 21 have both the controls and handles alongside one another, either at the side or top. One way to reconfigure is to move the control panel and display. Two possibilities are shown in Figure 22. Alternatively you might separate the control panel and display.

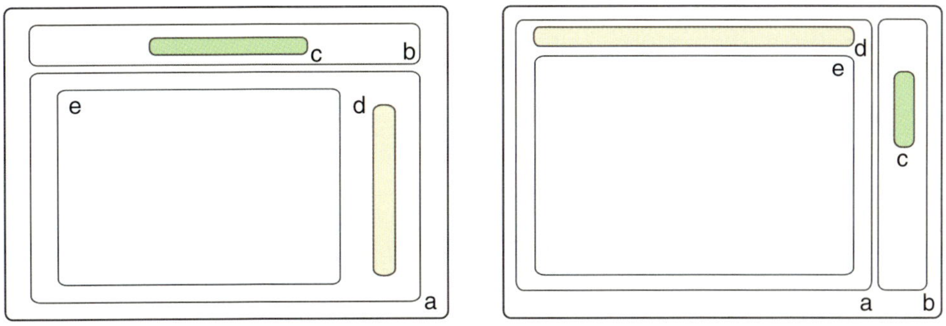

Figure 22 Two further configurations of microwave oven fronts

(b) Though your sketching was probably rough, it should show the shapes and sizes of each main part. For example, you might want to give the oven a playful appearance by using circular or oval shapes for the window and display.

Creating a layout includes decisions on configuration and components but each component has many properties in addition to shape and size. For example, the material it is made from, its colour and its texture are all critical features of a component. Different properties influence one another. For example, the type of material will affect the size and shape of a component – a stronger material can be used in thinner sections or different shapes. Another example of how one property influences another is how shape and configuration can be emphasised by colour.

(a) Ericsson phone (1975)

(b) Mobile phone (2003)

Figure 23 Configuration and colour

With the Ericsson telephone in Figure 23 the configuration of the main parts is emphasised by the black and red contrast. The buttons and handset are black while the main body is red. A more recent mobile phone, although smaller, also uses contrasting textures, colours and shapes to pick out and differentiate the main parts.

Many component properties all need to come together in a delicate balance to make the product look good, meet technical specifications and satisfy user needs. These are all aspects of how a product works – how it functions – there is more about this later in the block.

1.6.3 Sustainability issues influence product layout

A product affects many people during its life, not only the direct users. For example, raw material extraction and manufacture may deplete natural resources or cause pollution. These are environmental impacts of a product but there may also be significant social and economic impacts. The sustainability of a design is determined by all these impacts at the five main phases in the life cycle of a product (see Block 2). As a reminder, the five main phases are:

1 raw materials extraction;

2 manufacture – including materials processing and transportation;

3 distribution of finished product;

4 use;

5 disposal.

I concentrate here on the environmental aspects of sustainability, and in particular the materials and energy consumed during product life. The decisions made in design affect the resources used in each of the phases of the life cycle. Materials and energy are used to make the product, and energy is consumed in use. When the product reaches the end of its useful life the materials may be recycled.

However, there are many reasons why recycling is currently not as widely embraced as perhaps it should be. Partly, individuals base their purchasing decisions on product use, not product disposal, partly, many consumers do not think recycling is important. Also, individuals are rarely prepared to bear the costs of recycling where there is no immediate benefit to themselves. These are barriers to recycling and are being addressed by legislation, particularly in the European Union.

The environmental aspects of sustainability are complex. For example, increased recycling is not always accompanied by reduced energy consumption. In Europe cars are required to satisfy increasingly strict regulations on recycling and efficiency. However, the number of cars and their engine sizes continues to rise through market demand created by user preferences.

Layout design can have particular consequences for disposal and recycling. Being able to separate components made from different materials allows separate recycling. The way parts are joined together will determine how easily they can be separated and the number of parts will affect the costs of disassembly. For example, the front of a microwave may use many different materials: glass for the window, coated metal for the door, a plastic handle, a liquid crystal display, plastic buttons, and so on. For effective recycling all these have to be disassembled. The decisions taken in creating the layout all have a considerable influence on the environmental impacts of the final product.

The likely environmental impacts are important in determining the sustainability of a product. In designing it is too late just to evaluate the final product. Potential environmental impacts need to be evaluated at each stage of design.

SAQ 3

Think about the life cycle of a washing machine. Identify three different groups of people who might be affected by raw material extraction, manufacture, distribution, use and disposal.

1.7 Layout for a complex product

Exercise 10 below reviews the four aspects of a layout for a complex product: parts, arrangement, connections and form. Previously you have seen a layout design for an electric drill (Figure 5) and Exercise 10 uses a similar product, a hairdryer. Basically a hairdryer sucks in air at the back; a fan that is turned by a motor pushes the air past a heating element and out through the nozzle at the front. An on-off switch is on the handle and there may be controls for airflow and temperature. Figure 24 is a drawing of the configuration of the main parts of a typical hairdryer. The details of the motor, the switch and the heating element are not shown.

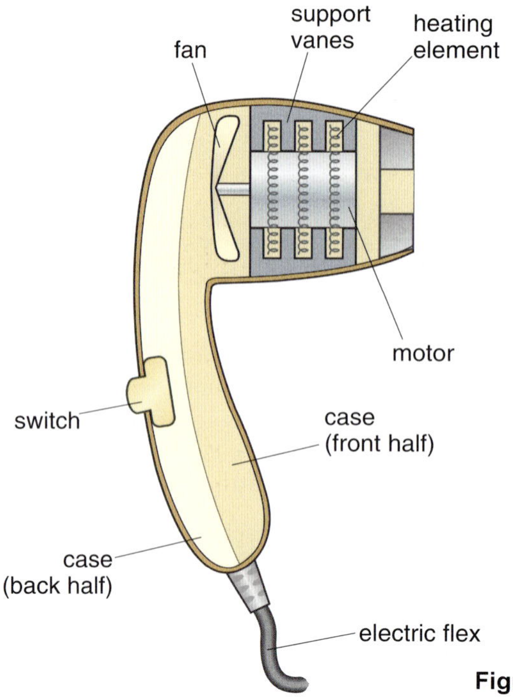

Figure 24 Main parts of a hairdryer

Exercise 10 Parts, arrangement, connections and form

Look at Figure 24 and describe in words the main parts, the arrangement, the connections and the form of the hairdryer. It may help to look at an actual hairdryer. Don't try to take a hairdryer apart.

Discussion

You can do this by listing the parts and against each one state the parts it is connected to and how it is connected. On the hairdryer that I looked at the main surrounding case of the hairdryer was in two parts – front and back. The interior surface of the case acted to hold all the other components in place. The heating element was supported on several vanes. These vanes also supported the motor. The subassembly of vanes, heating element and motor was fitted into the front half of the casing. The switch was fitted into the rear part of the casing.

The connections are a bit more tricky to describe. Electrical connections on the side of the motor are not shown. The electrical connections are made to the switch and to the controls for the motor and heating element. Finally the fan is connected to the shaft of the motor, perhaps with a spot of glue. The two halves of the case fit together with a screw and several push-fit mating surfaces on each half – these align the two halves, which click together on assembly. The screw is a safety feature and usually has a non-standard head so that it cannot be undone with an ordinary screwdriver.

Key points of Section 1

Developing a concept into a layout involves four stages.

1 Parts – identification of parts.

2 Arrangements – exploration of how the parts can be arranged.

3 Connections – specification of how parts are joined together.

4 Forms – specification of shapes, sizes and other properties of parts.

Stages 1 and 2 are referred to as configuration design, and stages 3 and 4 as component design.

Creating a layout is an iterative, not a sequential, process. Generating, evaluating and selecting possibilities takes place at each of the four stages.

The process of moving from a concept to a layout inevitably involves designers in a consideration of materials.

Published standards and specimen designs are useful in generating possibilities. Standards are essentially procedures for ensuring performance. Specimens are frequently used designs and may be modified to suit particular requirements.

Layout design can help in the effective reuse and recycling of products by ensuring they can be disassembled and the parts separated easily.

A designer needs to be thinking about the whole life cycle of the product right from the start of designing because decisions taken at the early stages of design can have a tremendous effect on the way the design is manufactured, what materials are used, energy consumption and how the product is disposed of or recycled.

Learning outcomes: 1.1, 1.2, 2.1 and 4.1

2 Configurations

In Section 1 I said there were four stages of developing a layout from a concept. In stage 1 the main parts are identified, and in stage 2 the main parts are arranged to produce a *configuration*. Stages 1 and 2 together make up *configuration design*.

Identifying the main parts in stage 1 may not always be as straightforward as it sounds. For example, a fork might be made from a single piece of stainless steel, but it still has parts in a manner of speaking – a handle and prongs (see Figure 25a). To make a distinction I call the handle and prongs features of the fork rather than parts because they belong to the same 'continuous' piece of material.

Sometimes parts are grouped together to form a larger part that has a clearly defined function. An individual bristle on a toothbrush is not really a significant part and neither is one cluster of bristles because by themselves they do not have a clearly defined function. However, the whole head of the toothbrush with its array of clusters of bristles has a clear function and is considered to be a part. The plastic handle of the toothbrush is also a part. The handle may itself have several smaller parts – such as embedded materials for gripping. This may remind you of the electric drill you saw in the previous section. The main parts of the drill all contain smaller parts, but during configuration design the designer concentrates on the main parts. So the toothbrush has two main parts: the handle and the bristle-cluster array forming the head (Figure 25b).

(a) Spoon, fork and knife

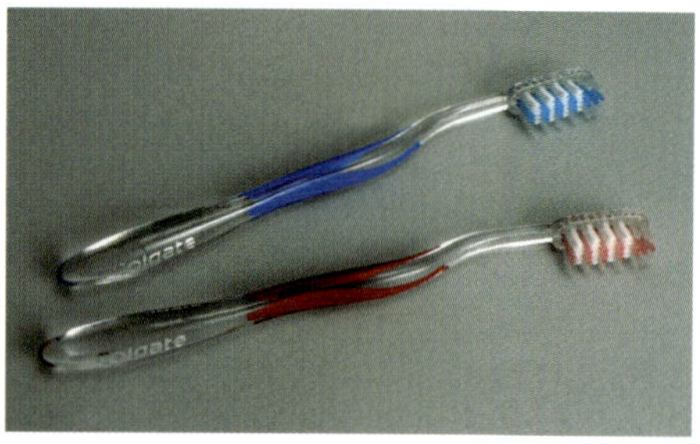

(b) Toothbrushes

Figure 25

In the previous section you considered the configuration of furniture in a room, the parts of a coffee table and the assembly of a hairdryer. A configuration is described by a model, often a drawing, showing where the parts are located.

An important function of this drawing is to show how the parts are located relative to one another. For the coffee table example, Figure 26 shows the relations of the top and legs as well as between shelf and legs. It is the relations among the parts that are important. The possible configurations of legs on the coffee table in Figure 8 reveal different relations between the legs. The pairs of legs could be arranged directly opposite each other, offset, or at right angles. The arrangement of features (prongs and handle) of a fork might be described as 'next to'. Furthermore if the fork is made in a single piece of stainless steel, the parts merge into one another. It is not always clear where one part ends and the next begins.

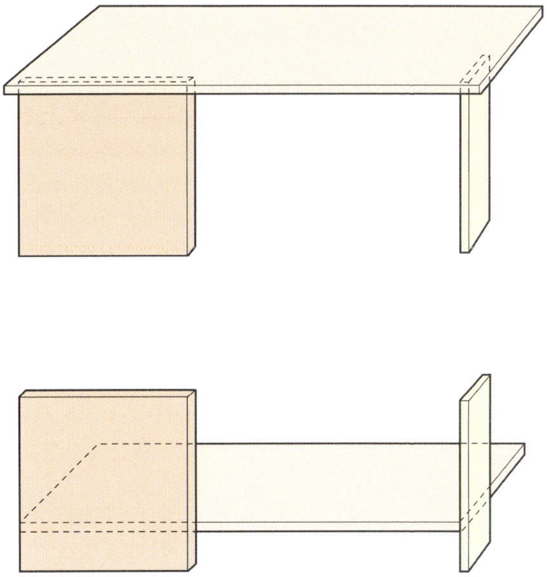

Figure 26 Relations between parts in the coffee table – between legs and top, and between legs and shelf

Exploring and generating configurations requires designers to have good skills in visual thinking. These skills are also needed for creating a layout. As well as needing skills in visual thinking, designers need skill in spatial thinking so they can generate different configurations and define arrangements between parts. Experienced designers have to integrate these all the time – it is part of the habit, pattern and routine of their jobs.

W 2

The development of visual and spatial thinking skills is supported by the *Shapes and Space Workbook*, to which you are referred at various points in Block 4. More information on using the *Shapes and Space Workbook* can be found in the *Block 4 Supplement*.

2.1 Describing a configuration

2.1.1 Simple relations between static parts

A configuration is described in terms of parts and the relations between parts. I will illustrate this by looking at several different designs including brick walls, house roof structures, bicycle parts and house floor plans.

(a) One brick thick

(b) Two bricks thick

Figure 27 Brick patterns in walls

A brick wall has bricks laid in a pattern. The parts, the bricks, are roughly the same size and shape. For strength, the relations between the bricks are such that any one brick shares faces with others. The relations correspond to physical connections between the bricks made by mortar, which bonds the faces to each other. In Figure 27 two different brick wall patterns are shown. In pattern (a) one brick is related to other bricks by sharing short faces end to end, and by overlapping on the top or bottom face.

Pattern (a) is a thin wall just one brick thick. Each brick is related directly to six other bricks, two at the ends, two at the top and two at the bottom. In the thicker wall (b) bricks are related to either eight or nine other bricks.

Exercise 11 Brick connections

Look at pattern (b) in Figure 27 and identify which bricks have eight and which have nine connections to other bricks. Then list the types of relation between bricks in this wall.

Discussion

Look at bricks in the body of the wall, not ones on top of the wall. Those bricks that show their small end at the surface of the wall have eight connections while those showing their sides at the surface of the wall have nine connections. Figure 28 isolates the connections (this is a simplification and the connections do not match Figure 27(b) exactly).

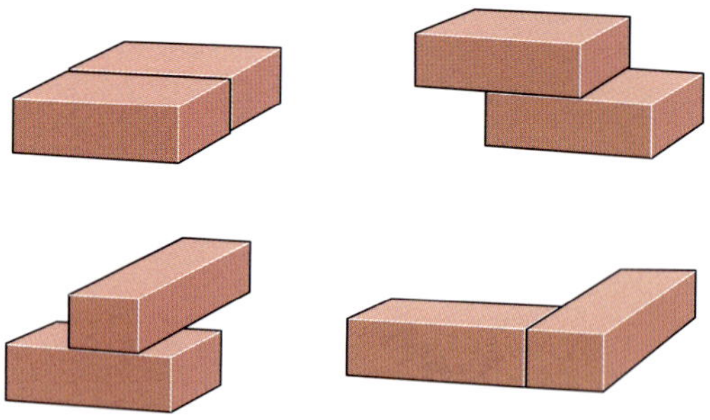

Figure 28 Relations between bricks

Figure 28 shows bricks are related by:

- sharing long side faces;
- overlapping top and bottom faces with aligned bricks;
- overlapping top and bottom faces with bricks at right angles;
- sharing long and short side faces with bricks at right angles.

Notice that pictures can be better than words at describing the different relations between parts.

The relations among the bricks give the wall its structure – they make it strong enough to withstand loads and disturbance. New parts in a wall would introduce new relations, an example would be a lintel above a window.

Brick walls have a large number of small identical parts that can be made into many different shapes and sizes. Other structures in a building use fewer but larger parts. The pitched roof of a house is constructed from lengths of timber joined in a truss that forms a fairly large structure.

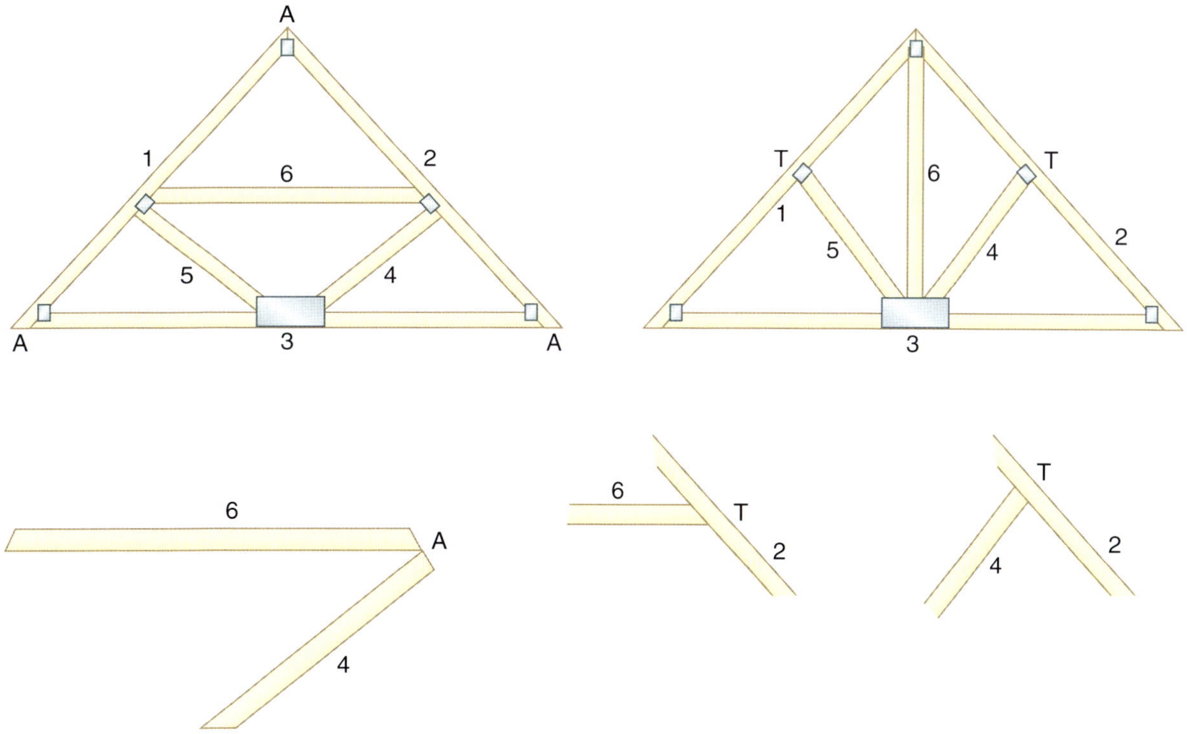

Figure 29 Roof trusses. At the top-left is the type with a short horizontal beam; at top-right is the type with a long vertical beam; and at bottom is how parts meet at the joints.

There are two different relations in Figure 29. First, beams can be end to end at an angle. In the top-left, truss beams 1, 2 and 3 are connected end to end in this way. These connections are marked 'A'. Second, beams are end to centre, again at an angle. In the top-left truss the connections between 4 and 2 and between 6 and 2 are of this type. I call the first type A joints and the second T joints. In the top-right truss part 6 is connected to both 1 and 2 by A joints. Part 6 is connected to 3 by a T joint. Another aspect of these connections between beams in trusses is that at a single joint there may be some connections that are type A and others that are type T. The joint where beams 2, 4, 6 join in the top-left truss has beams 2 and 4 connected by a T joint. However, beams 4 and 6 are connected by an A joint.

Notice the two different configurations of truss are both made up from triangles because triangles provide rigid structures. Triangles present three-way connections in which each of the parts is related to the other two. These triangular relations among parts are a feature of structures that are designed to be strong and rigid.

With different possible configurations of roof truss available a designer decides which configuration to use, the dimensions of timbers and how to join the parts. The decision will probably be influenced by loads on the roof, the span, the pitch, and the need for space inside the roof.

SAQ 4

Identify two examples of structural design in which triangles of parts are used extensively.

2.1.2 Configuration with mechanical and electrical parts

Now consider relations between parts in a more complex product. Look at the picture of an electric drill in Figure 30, and have a look at the cut-away you have already seen. The main parts of an electric drill are the two halves of the casing, motor and gearbox (considered as one part), shaft from the motor, bearing, chuck, controls and battery pack. The main parts each have a clear function, and some of the main parts have many smaller parts. For example, the motor has static and rotating windings on laminated iron cores. The two-part outer casing is referred to as a clamshell because the two halves come together like a shell.

Figure 30 Portable electric drill and cut-away

The main parts of this electric drill are related in several different ways. The motor is fixed in the casing, the shaft rotates in the motor and the chuck is fixed rigidly to the shaft. These are all mechanical connections. The battery, controls and motor are connected electrically.

2.1.3 Network diagram for representing configuration

It would be useful to have some way of representing these connections. One way is to draw a diagram that reveals the network of connections. Such a network diagram uses a small circle or dot to represent a part and a line joining two parts to represent a connection. The line can be labelled with the type of connection (see Figure 31).

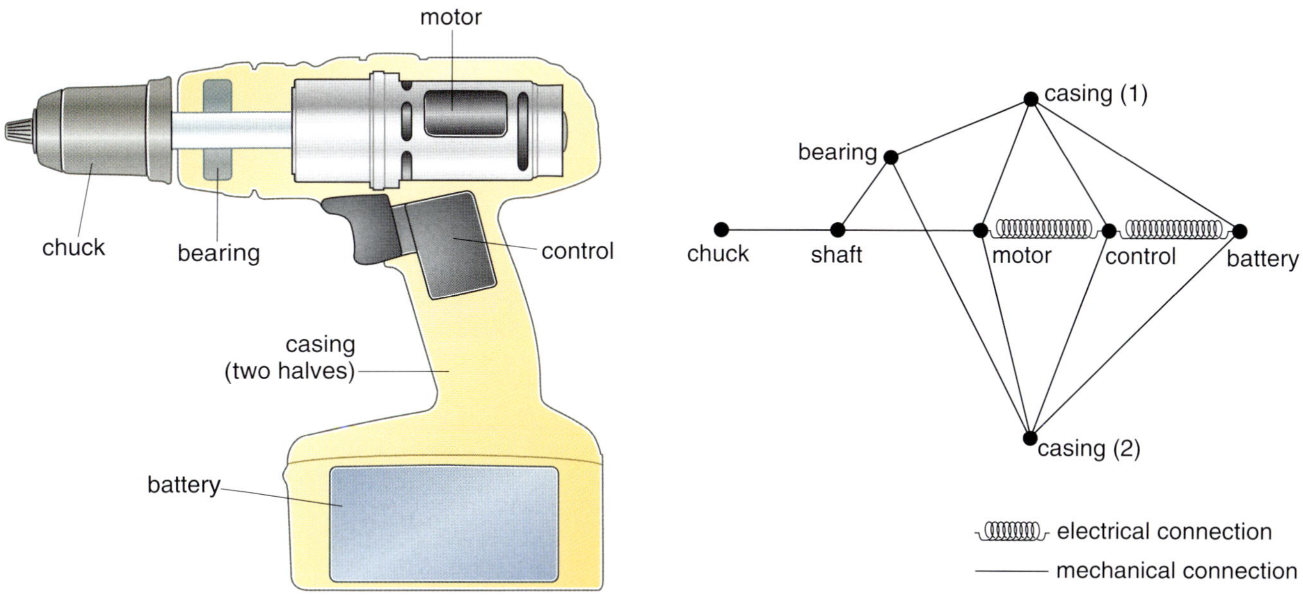

Figure 31 Main parts of a portable electric drill and network of connections

Exercise 12 Network connections

Construct a network of connections for the two roof trusses shown in Figure 29. I suggest you first number each of the timbers in the truss as in the figure and indicate the type of each connection by labelling with A or T.

Discussion

In each truss there are six parts – three for the outside triangle and three internal parts. As mentioned above there are two types of connection: the first joins two beams at their ends, the second joins the end of one beam to the middle of another beam (see Figure 32).

Figure 32 Networks of connections between parts of roof trusses

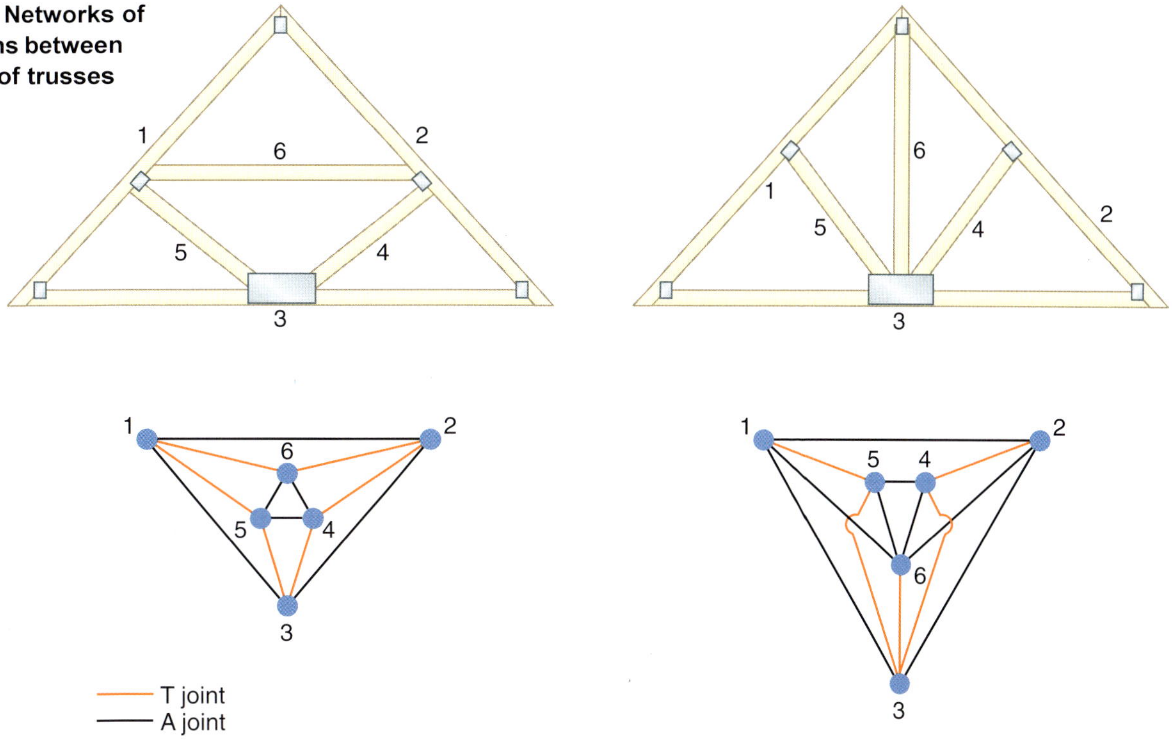

Some of the connections between parts in the electric drill must be present for the product to work – such as the mechanical connection between motor and chuck or the electrical connection between battery and motor. Other connections may be less definite. For example, the battery could be at the base of the handle as in the example shown or it could be mounted on the top of the casing. There may be reasons not to do the latter, such as the imbalance of the drill, but it is a possible configuration.

2.1.4 Configuration with static and moving parts

Up until now I've considered the drill as a configuration of static parts but part of the drill rotates. The rotation is allowed by a moving connection inside the bearings of the motor.

Other products also have moving connections. Recall the Strida bicycle in Figure 2. There is a sequence of connections: pedal, crank, chainwheel, toothed belt and wheel. The motion of each part determines the motion of the next part in the sequence. Some of the connections are between moving parts. The pedal rotates relative to the crank and the belt moves relative to the chainwheel. Figure 33 is a way to show a network of fixed and moving connections.

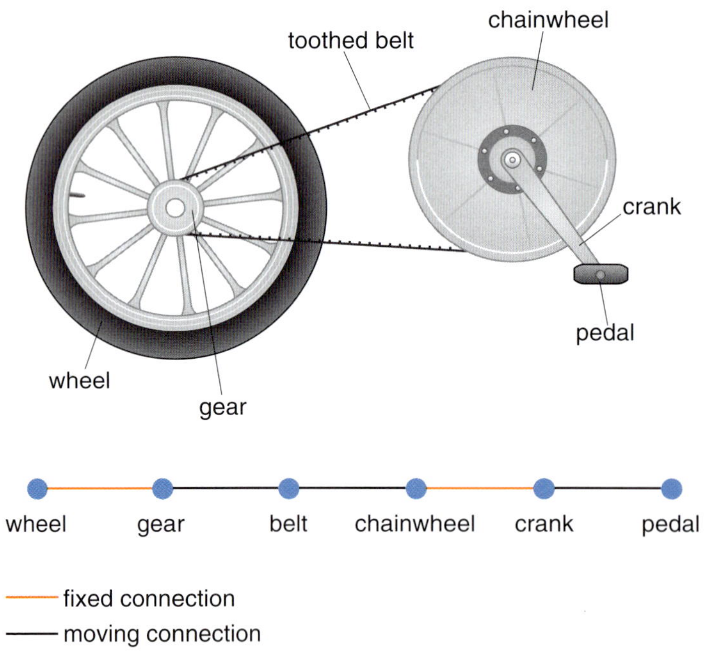

Figure 33 Fixed and moving connections in parts of a bicycle

2.1.5 Configuration with less distinct parts and connections

So far connections have mostly been between parts that are next to one another – direct physical connections. But designs can have different kinds of connections. Your computer has parts – DVD and hard disk connected by cables (as special connecting parts), and peripherals like a printer that may be connected by a cable or a wireless link. The 'remote' computer connections communicate signals between parts. I could argue that a mechanical design is similar. For example, in the Strida bicycle the rotation of the pedals is communicated by the teeth on the belt to the wheel.

Are there other types of relations between parts? I have looked at static, moving and remote connections between separate and distinct parts but sometimes the parts are not physically distinct. Recall from the introduction to this section that what can be referred to as parts may instead be distinct features on a single piece of material, as in knives and forks.

Another example is a cardboard box. Most cardboard package designs are cut from a sheet of card as a single piece (you made some yourself in the early part of the *Modelling Workbook*). The features of the cardboard sheet can be identified as top, sides and bottom of the box and will usually be printed. The overall two-dimensional shape is then cut out, folded and glued to make the box. The connections between the features of such a box are of two types – a fold line in the material and a glued edge (Figure 34).

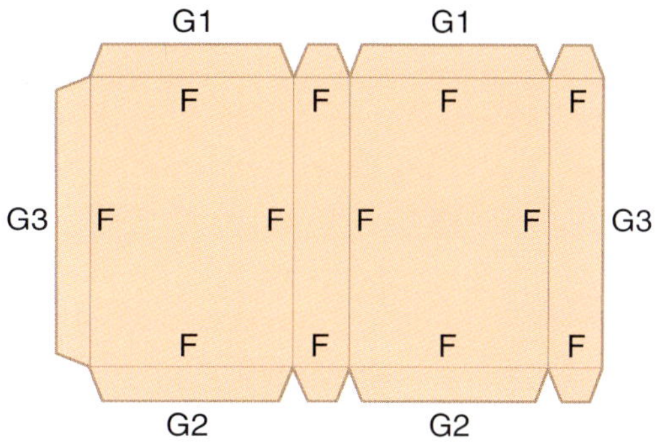

F–fold
G1,G2,G3–glued connections

Figure 34 Box cut-out showing positions of folds and glued connections

2.1.6 What parts should be considered?

I will finish this discussion of parts and connections with an example in which the relations are a little harder to analyse. Look at the plan for the ground floor of a house with five rooms (Figure 35). Walls are the continuous straight lines in the plan. There are four walls around the house and a wall right down the middle of the house from front to back.

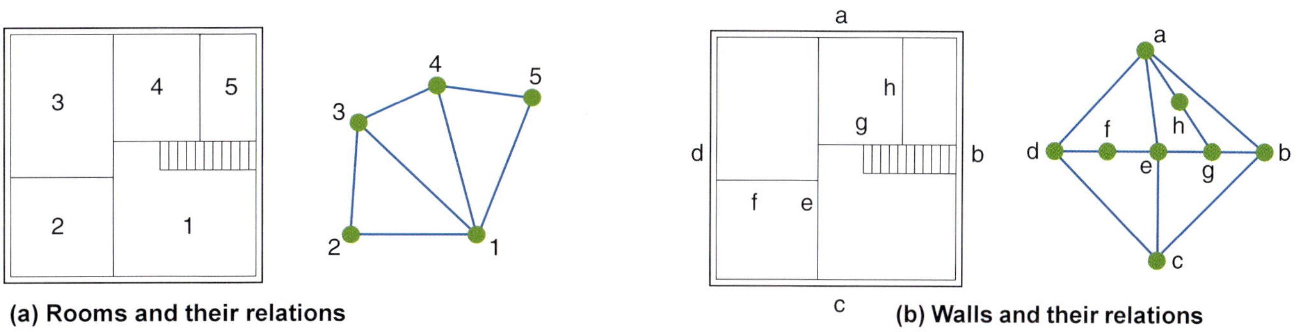

(a) Rooms and their relations

(b) Walls and their relations

Figure 35 Five-room house plan

The rooms are numbered 1 for the hall and staircase, 2 for the living room, 3 for the dining room, 4 for the kitchen, and 5 for the utility room. The walls are labelled with the letters 'a' to 'h'. To be connected in this network, rooms do not necessarily need to have a door between them – just a length of shared wall. For walls to be connected they only need to abut against another wall.

When considering relations between parts you need to decide which are the parts for consideration. Figures 35(a) and 35(b) are the same house plan but (a) shows rooms as connected parts, whereas (b) shows walls as connected parts.

This example demonstrates that the configuration of a design depends on what parts and connections are chosen to represent a design.

Exercise 13 House plan connections

Describe by two networks the connections between walls and the connections between rooms in the house plan in Figure 36. What do you notice about the networks in Figure 36 and those in Figure 35?

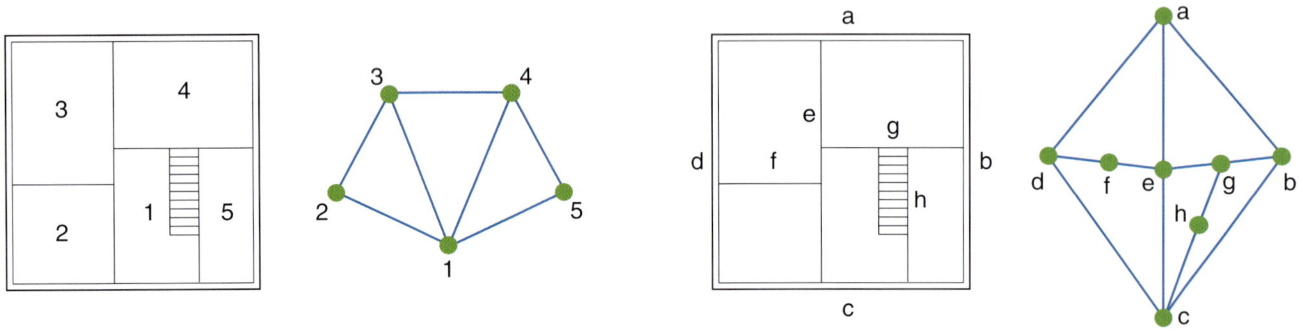

Figure 36 Another five-room house plan

Discussion

The networks for the room connections are the same. This means the configurations are the same even though the shapes and sizes of the rooms may be different. However, the networks of the wall connections are different.

You have seen descriptions of existing designs as configurations of parts. But this is only one side of understanding design. A designer's aim is not primarily to analyse existing designs but rather to create new configurations and therefore new designs.

W 2

This would be a good point to begin the *Shapes and Space Workbook*. I suggest you read Section 1, titled 'Two- and three-dimensional shapes'.

2.2 Configurations of repeating parts

I will now examine in more detail the ways that configurations are constructed from parts. There will be constraints from the design brief that will limit possible configurations. Alongside these external constraints are limitations imposed by the types of parts and connections included in the configuration. For example, with the configurations of bricks (Figure 27), the parts were limited to a single brick size. The connections between parts were also limited by the regular overlapping pattern of bricks for a strong wall. These constraints result in a small number of viable configurations of bricks in a wall.

Connections and relations between parts (bricks) were shown in Figure 27. These relations are used to create the patterns of repeating bricks in a wall. The way they repeat is quite complex and three-dimensional. To explain repeating patterns in a simpler way I will examine two-dimensional repeating parts.

Household items such as ceramics and printed fabrics often display repeated parts. The repetition is generated by moving a part – a motif as it is called in a two-dimensional pattern. The result is a regular pattern. I think it is best to distinguish these different motif movements as repetition by rotation, reflection or translation (see Figure 37). Sometimes these repetitions are called symmetry operations because they can generate symmetrical patterns, as you shall see shortly.

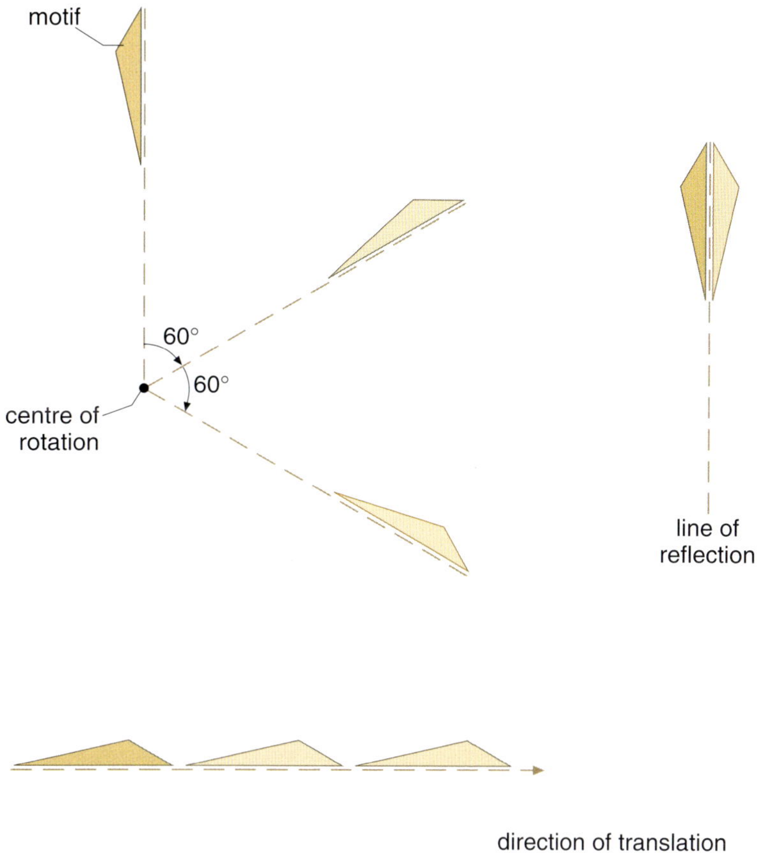

Figure 37 Motifs repeated by rotation, reflection and translation

Figure 38 Patterned plates with a motif repeated by rotation

For a rotation the motif is repeated by turning it about a centre of rotation. The pattern is built up by a sequence of repetitions. Two plates are shown in Figure 38. The lower plate has decoration around the rim created by rotating a motif about the centre of the plate. The motif is rotated by 15 degrees in each repetition. The result has 24 copies of the motif around the rim of the plate. The upper plate, with a quite different motif, has a similar repetition, but the pattern is now concentrated at the centre of the plate.

For a reflection, the motif in Figure 37 is mirrored in a line. The reflection by itself does not repeat the motif any further. Another reflection just brings the motif back to its original position. In order to repeat indefinitely the motif needs to be translated, in other words, moved. The translation repeats by moving the motif but in all other respects keeping it the same. Applying a reflection and a translation gives the pattern shown in Figure 39(a). The sequence used here is reflection–translation–reflection and so on. In principle this pattern goes on indefinitely. Applying a rotation and a reflection in the sequence reflection–rotation–reflection and so on gives the pattern in Figure 39(b). This pattern comes back to meet itself provided multiples of the rotation add up to 360 degrees.

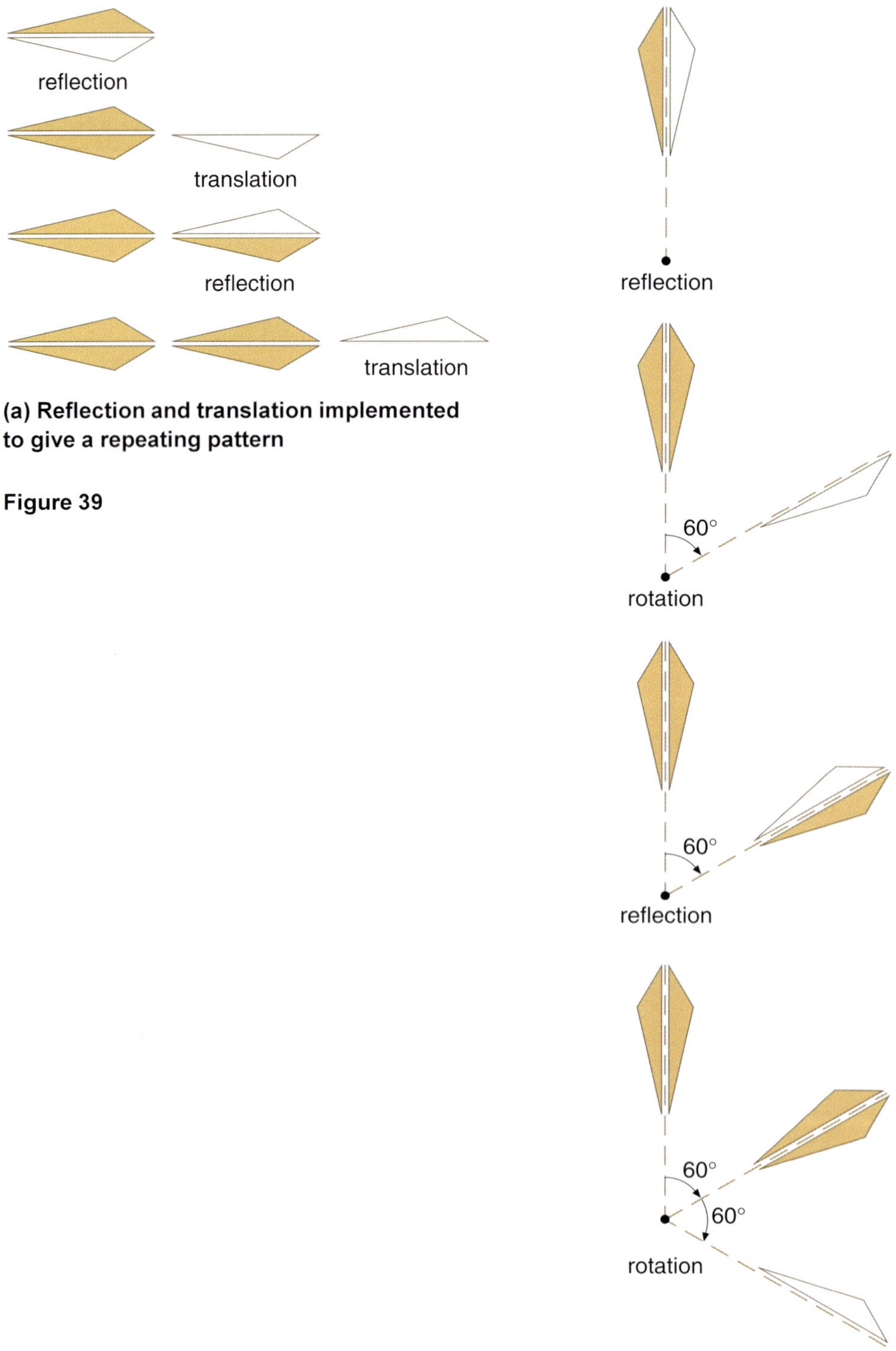

(a) Reflection and translation implemented to give a repeating pattern

Figure 39

(b) Reflection and rotation applied to give repeating pattern

Exercise 14 Creating pattern

Each repetition in Figure 37 corresponds to a single rotation, reflection or translation. The repetition in Figure 40 is more complicated and comes from a combination of translation and reflection. What pattern is produced by this repetition?

Figure 40 Repetition that uses both translation and reflection

Discussion

The rule creates a pattern that repeats along a line but zigzags either side of the line, resulting in the pattern in Figure 41.

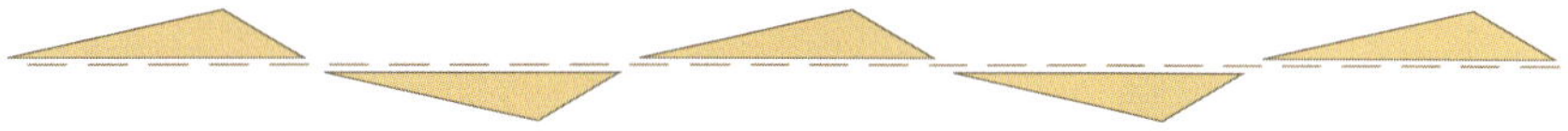

Figure 41 Repeating pattern by applying repetition in Figure 40

A motif might move and change colour. It is then not strictly a repeat because the colour changes, but the shape is repeated. Figure 42(a) is a translation of both shape and colour. Figure 42(b) is a translation of shape combined with colour change.

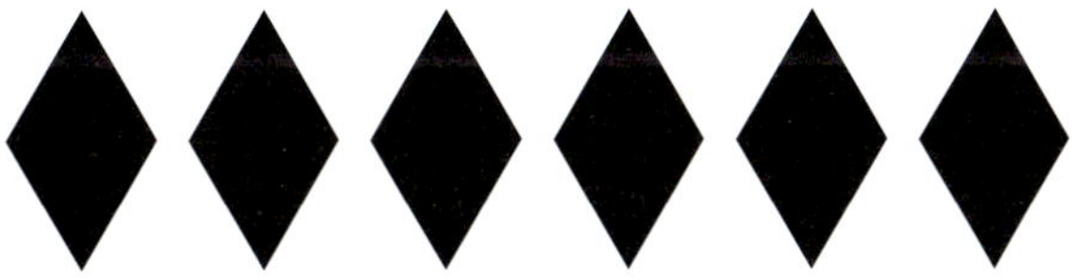

(a) With no change of colour

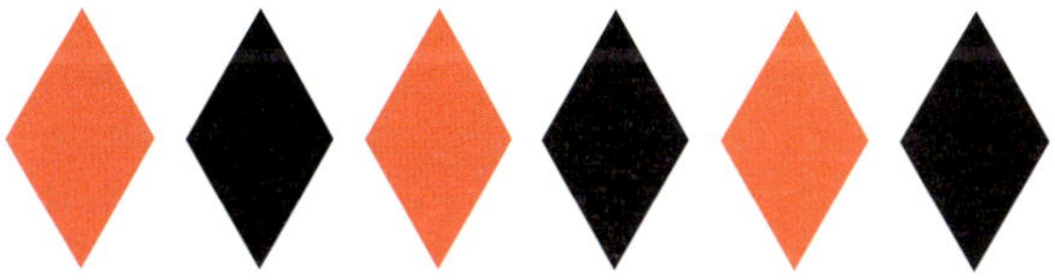

(b) With change of colour

Figure 42 Motif translating

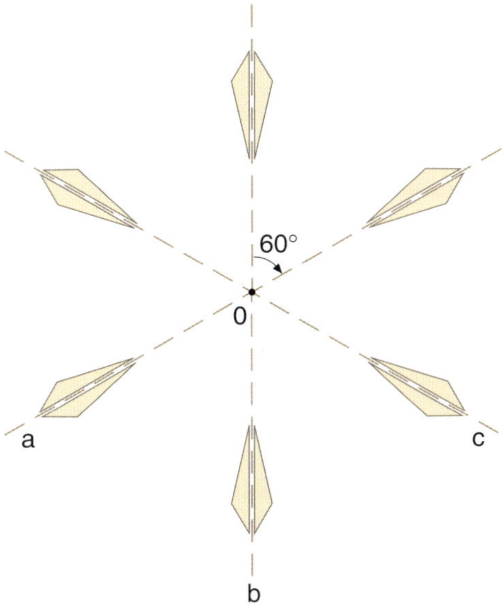

Figure 43 Pattern showing rotational and reflection symmetry

The patterns created by repetitions are regular. In some cases this regularity is strong, so it is said the pattern is symmetrical. Take another look at the plates in Figure 38. The repetition 'clocks' around until it returns to its start. Look at the whole pattern. Turn it through 15 degrees. The whole pattern looks the same after rotation – it is rotationally symmetric (see Figure 43). A pattern is symmetrical if there is a way to move the whole pattern – by rotation, reflection, translation or some combination – so that it stays the same after its move as it was before. This type of move of the whole pattern is called a symmetry of the pattern.

SAQ 5

For the two patterns of lozenges shown in Figure 44, find three symmetries for the first pattern and one symmetry for the second.

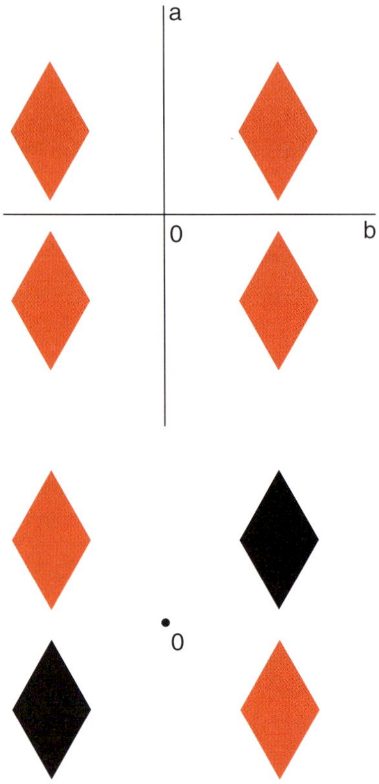

Figure 44 Patterns of lozenges

Figure 45 Sunburst clock is nearly symmetrical with 12-fold rotational symmetry

There is plenty of repetition of shapes in many ordinary products. Symmetry often comes from relatively simple repetitions. A car body is nearly symmetrical because it is almost the same on right and left. Many products are nearly symmetrical because many but not all of the parts are repeated.

Breaking symmetry slightly can make a design appear interesting while maintaining the strong rhythm of repeating elements. For example, a clock's digits break exact symmetry – as of course do the hands that stand out against the static background symmetry. The clock in Figure 45 is nearly symmetrical. It has no digits and just a double mark to indicate the top of the clock.

W 2

This would be a good point to read Section 2 'Organising elements on a plane' of the *Shapes and Space Workbook*. In addition to developing the concepts discussed here, it introduces the next two subsections of this block.

This would be a good point to do the activity on the DVD called 'Exploring configuration'.

This would be a good point to do the activity on the DVD called 'Symmetry and everyday objects'.

This would be a good point to watch the DVD video sequence 'Repeating patterns'. This shows examples of repeating patterns in terraces of houses.

2.3 Repeating patterns in housing layouts

Most objects are not exact repetitions of fixed-size parts. Parts with a specific shape are repeated but their size is allowed to vary. For example, in a house plan rooms are rectangular but of different sizes.

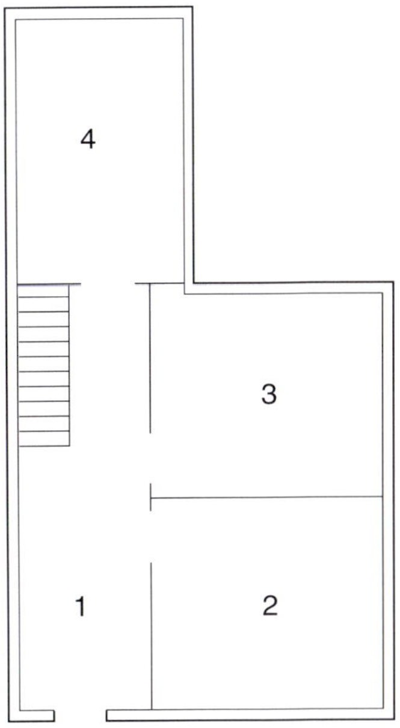

Figure 46 Typical terrace house L-plan

2.3.1 Terrace house configurations

First, I'll consider a terrace house built in the nineteenth century (see Figure 46). A typical ground floor plan has a front door giving onto a narrow hallway to stairs and access to a front and back room. At the rear is a kitchen. This deep and narrow plan has four main spaces and a staircase. Kitchen, stairs, sitting and dining rooms all need access onto the hallway. All rooms are next to the outside for light or access. Further constraints, which may have been imposed by local regulation or custom, concern access to the rear of the house, perhaps from a back lane.

There are few viable terrace house plans especially if the width of the front of the house giving onto the street is narrow. The plots on which these terrace houses were built were characteristically narrow and deep. There were good reasons for this. Perhaps developers wanted as many houses fronting a given length of street as possible, or to minimise the amount of land given over to street, or to reduce the length of service runs for gas, water and sewers along the streets.

These L-shape plans are repeated along the street. How can the L shape be repeated? Figure 47 shows three possible ways that neighbouring houses can be related: short side to short side (SS), long side to long side (LL), or short side to long side (SL or LS).

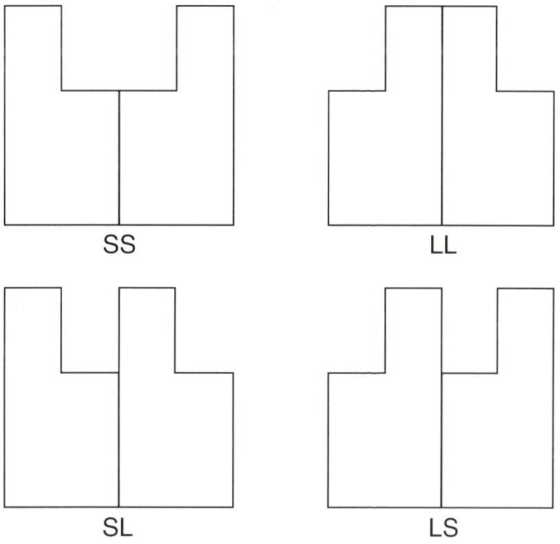

Figure 47 Relations of adjoining terrace houses

Usually houses in a terrace are arranged regularly in one of two ways (Figure 48). The neighbouring houses repeat either by translation or by a combination of a translation and a reflection. Notice in Figure 48 that one of the configurations has front doors paired together and in the other they are regularly spaced along the terrace. This is exactly what you can see in the DVD sequence on repeating patterns.

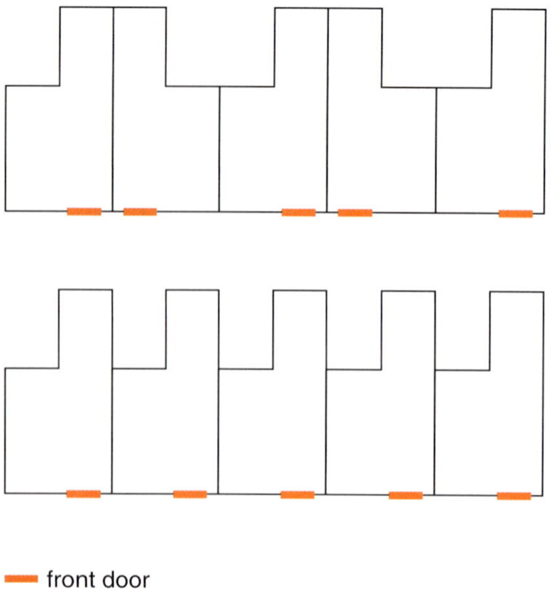

Figure 48 Alternative front door positions in terrace houses that are related SS+LL (top) and LS (bottom)

2.3.2 Semi-detached house configurations

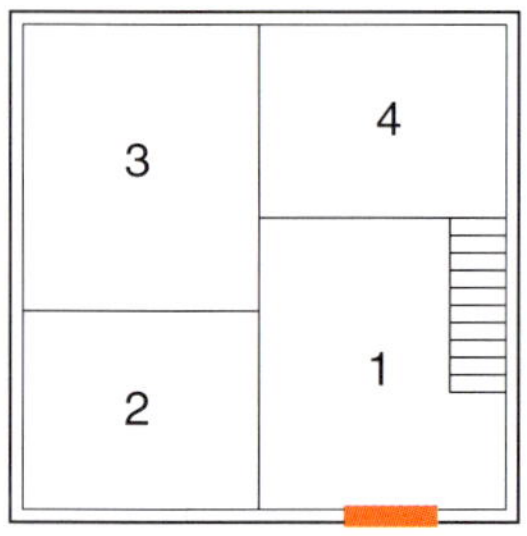

Figure 49 Semi-detached house plan

In the twentieth century, semi-detached houses became more widespread in English suburban developments. Typically the streets were longer than terraced streets and each plot was larger with a wider frontage. The semi-detached configuration allowed access and light to the side of the house. The same rooms were arranged in slightly different ways. The plan assumed a different overall shape – wider at the front and less deep from front to back. Houses became lighter inside. A possible ground floor plan is shown in Figure 49.

These houses were repeated by a reflection rule as well as a translation rule, which repeats them down the street. The translation moves the house further down the street than for terrace houses – it leaves a space between the houses. Two ways were possible: front doors were either together or far apart (Figure 50).

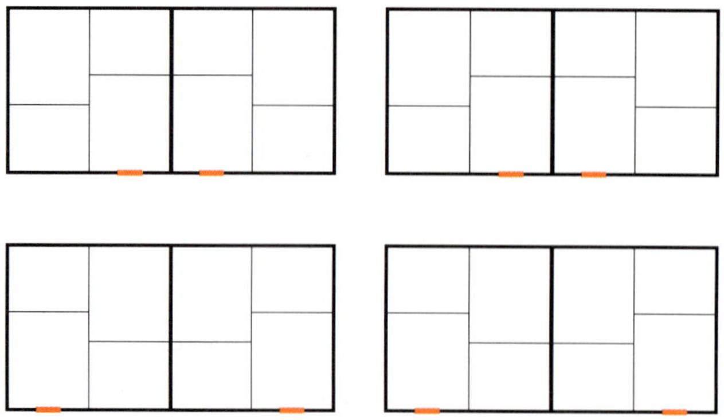

Figure 50 Two configurations of semi-detached houses repeated in a row with alternative front door positions

The lower option in Figure 50 may have given a sense of a separateness and was sometimes preferred. However, the upper option gives living rooms away from the dividing wall, although this necessitated separate chimney stacks for each house. As you will be aware, the pairs of houses may be repeated along straight streets, or around curves or crescents.

2.4 Configurations in products

You've seen how repeated parts create regular patterns in configurations. Now I'll look at how sets of different parts are configured. Without repetition the patterns are less regular.

2.4.1 Brands identified by configuration

Although individual products may not display repeated parts across a product range, there may be common parts and relations that identify the products of a particular company or even designer. Some parts may be used widely across many products and may even be used to indicate brand.

Brand may be indicated by a name or colour – perhaps by a logo for the name or a colour in the product or packaging. Logos for well-known brands such as Coca-Cola and Cadbury use colour, lettering and shapes and are valuable company assets that can become known worldwide. But brand is not just about packaging. Furnishing company IKEA is a brand denoting a type of design or lifestyle associated with products.

Some domestic products such as the Alessi kettle in Figure 51 have a strong and coherent identity through their distinctive geometric shapes or contrasting colours of elements. Similarly the Dyson brand is identifiable through the consistent application of geometric forms and bold colours. Look at various Dyson vacuum cleaners when you next have the chance. Look at how elements such as cylinders have been used as building blocks for the overall form (Figure 52). Look also at how various separate cylinders can be unified or separated with colour, texture, material, and so on. Handling parts consistently across a range of products can produce a visual brand identity.

Figure 51 Alessi kettle is identifiable because of the spheres and cones, shiny steel and contrasting colours

Figure 52 Dyson vacuum cleaner is identifiable because of the cylinders, colours and transparency

Exercise 15 Geometry in kettles and cleaners

(a) For the Alessi kettle identify the main parts and the relations between them.

(b) In the Dyson cleaner identify separate cylinders on the main part of the product and the spatial relations between them.

Discussion

(a) The main parts of the kettle are quite simple geometric shapes (Figure 53a). The body of the kettle is a cut-off cone. The lid completes the cone but is rounded and almost spherical. The knob on the lid is nearly a sphere on the apex of the cone. The spout is another cut-off cone attached to the body of the kettle at an angle. There are further decorative parts. These are all based on simple shapes but in contrasting colours and textures (the handle is not shiny like the main kettle). The handle is a circular arc broken on one side by the regular wavy line of the grip.

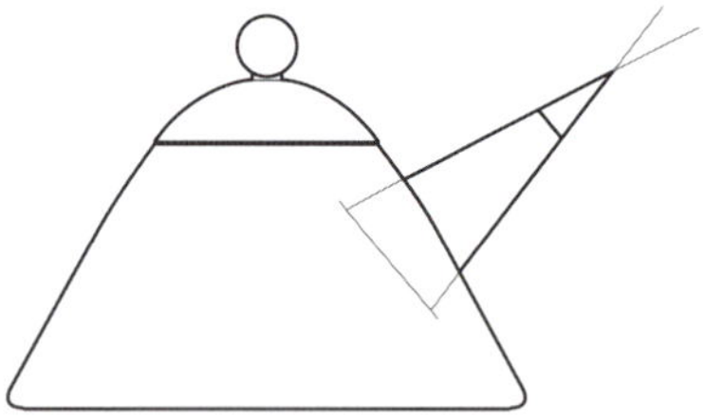

(a) Relations between cones, spheres and cylinders in Alessi kettle

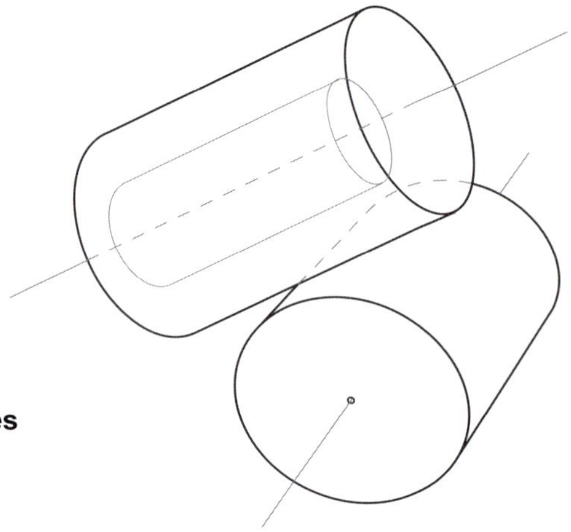

Figure 53

(b) Relations between cylinders in Dyson vacuum cleaner

(b) At least three cylinders can be identified. I noticed immediately the main transparent one, part of the cyclone inside and the wheels (Figure 53b). Relations include an alignment along an axis (for the transparent cylinder and the cyclone) and axes crossing at right angles.

2.4.2 Similar configurations

Configurations in products of a particular type are often similar. Cutlery is not usually designed as individual pieces but as a product range. Generally configurations of features are standard across all cutlery.

Products that are more complex also share configurations. Most cordless electric drills have the same configurations of parts. The underlying configuration of battery pack, motor, controls, handle, case and chuck stays the same across different drills. Styling and variations in shapes and sizes visually distinguish one product from another.

2.4.3 Exploring configurations

For really new products deciding on the configuration requires a look at all the possibilities. Take, for example, the Strida bicycle. In the patent application many possible configurations were illustrated to lay claim to the variants of the triangular frame, the different methods of unfolding the frame and driving the wheels as well as the positions of handlebars and seat. In Figure 54 dotted lines show possible ways of joining, and the figure numbered 'FIG.17.' in the patent drawing represents the configuration actually used, with chainwheel located on frame member C.

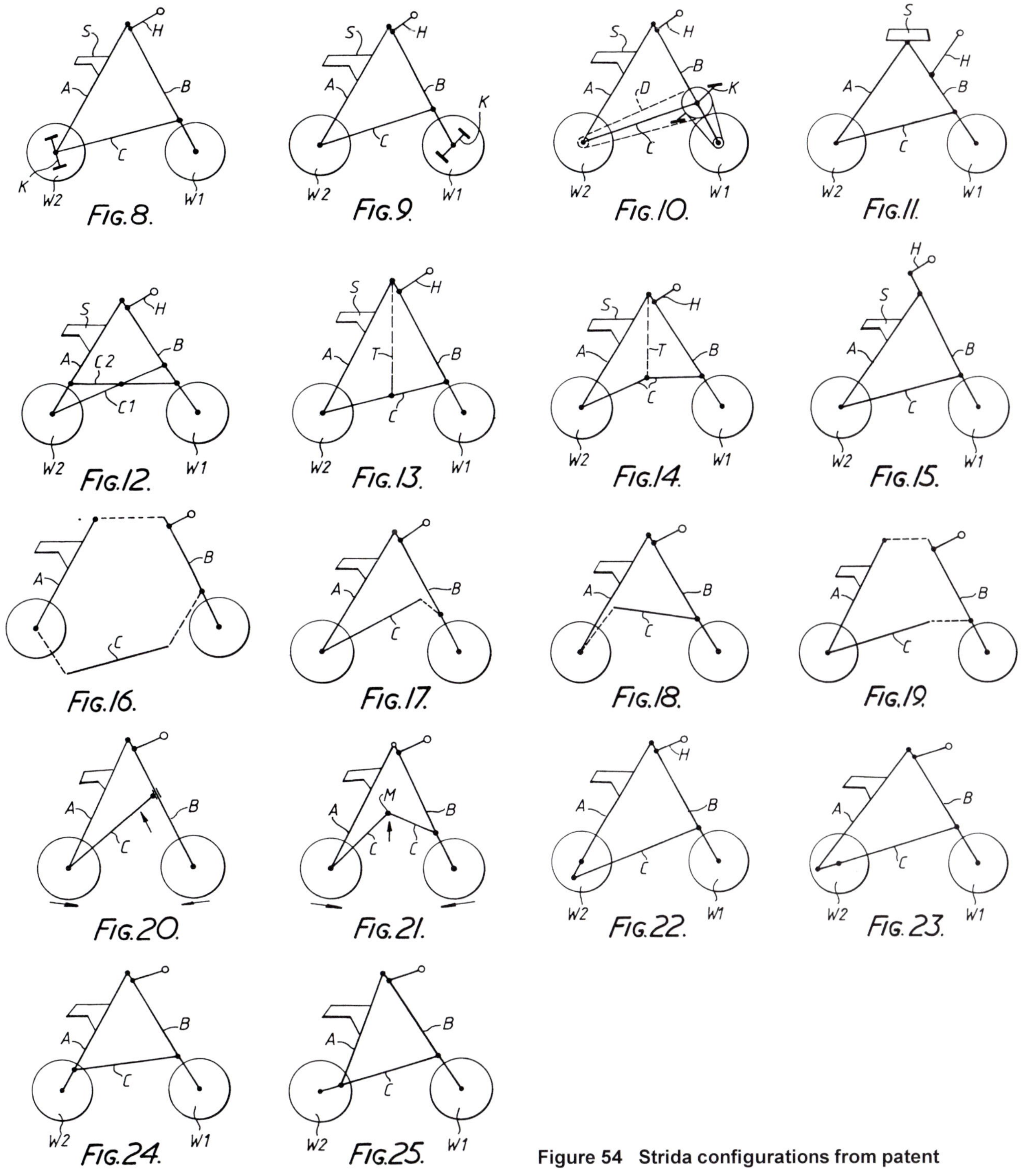

Figure 54 Strida configurations from patent

Figure 55 Strida with lower frame member undone

Relations between parts give useful functions. For example, when the toothed belt on the Strida bicycle engages with the teeth on the chainwheel the relation allows the transmission of force. The toothed belt and the toothed chainwheel work together – they are a pair of related parts with a particular function. The frame has three related parts in a triangular structure that is rigid. Recall the rigid triangular relations between parts in the examples of roof trusses. When one of the relations is disconnected the bicycle can fold (Figure 55).

Related parts in a triangular structure give a rigid configuration. One way to see how a triangle keeps parts still is to make a square from strips of card joined at the corners with drawing pins. The square can change shape into a parallelogram (middle shape in Figure 56). Putting another strip of card across a diagonal creates two triangles and stops this movement (bottom shape in Figure 56).

A row of triangles added together edge to edge is also rigid. This is the kind of structure you might find in a crane, a roof truss or a bridge (Figure 57).

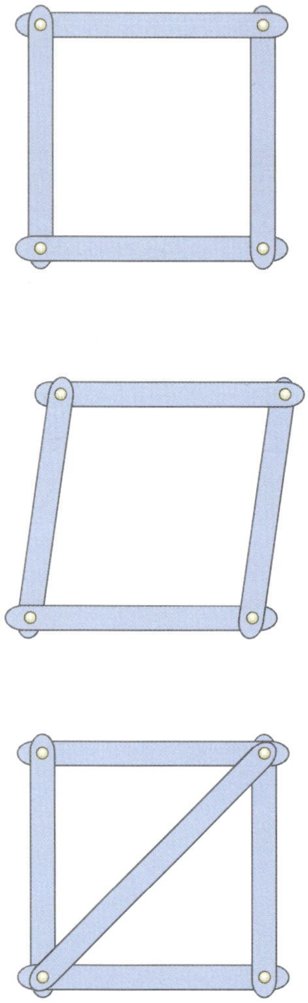

Figure 56 Four connected strips able to move (top and middle) and made rigid by diagonal (bottom)

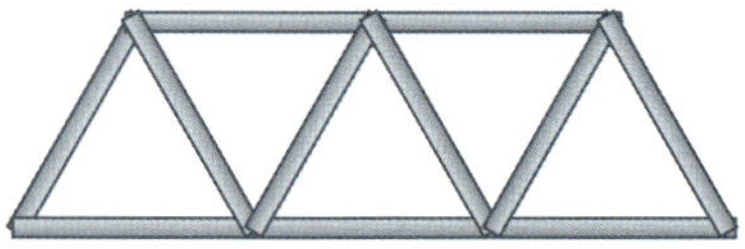

Figure 57 Adding triangles together gives a rigid configuration

2.4.4 Controlled motion – a mechanism

Configurations with connections that can move and are not rigid are called mechanisms. They transmit motion from an input to an output. The input and output motions may be quite different. Useful mechanisms convert motions that are easy to produce into desired motions – like the rotations from a motor in a sewing machine being converted into the up and down motion of the needle. Moving one part of a mechanism causes others to move. To explore this further I will examine how to configure parts to create one particular mechanism – a lifting hoist.

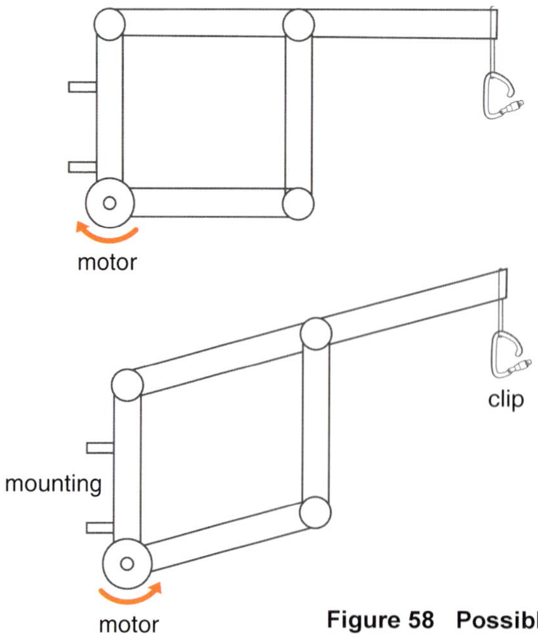

Figure 58 Possible configuration for a hoist

One possibility for creating a hoist is to construct a jointed parallelogram with an electric motor at one of the joints (Figure 58). However, this configuration does not work well. The turning force (torque) of the motor has to be large to lift a heavy load. In practice, other configurations are used to lift the load.

Look at the hoist for lifting people from a wheelchair into a car (Figure 59). Two additional rotations allow the person to swing in through the car door. These are not driven motions. Once the person is lifted out of the chair using the cradle, a helper swings the cradle in by hand through the door opening.

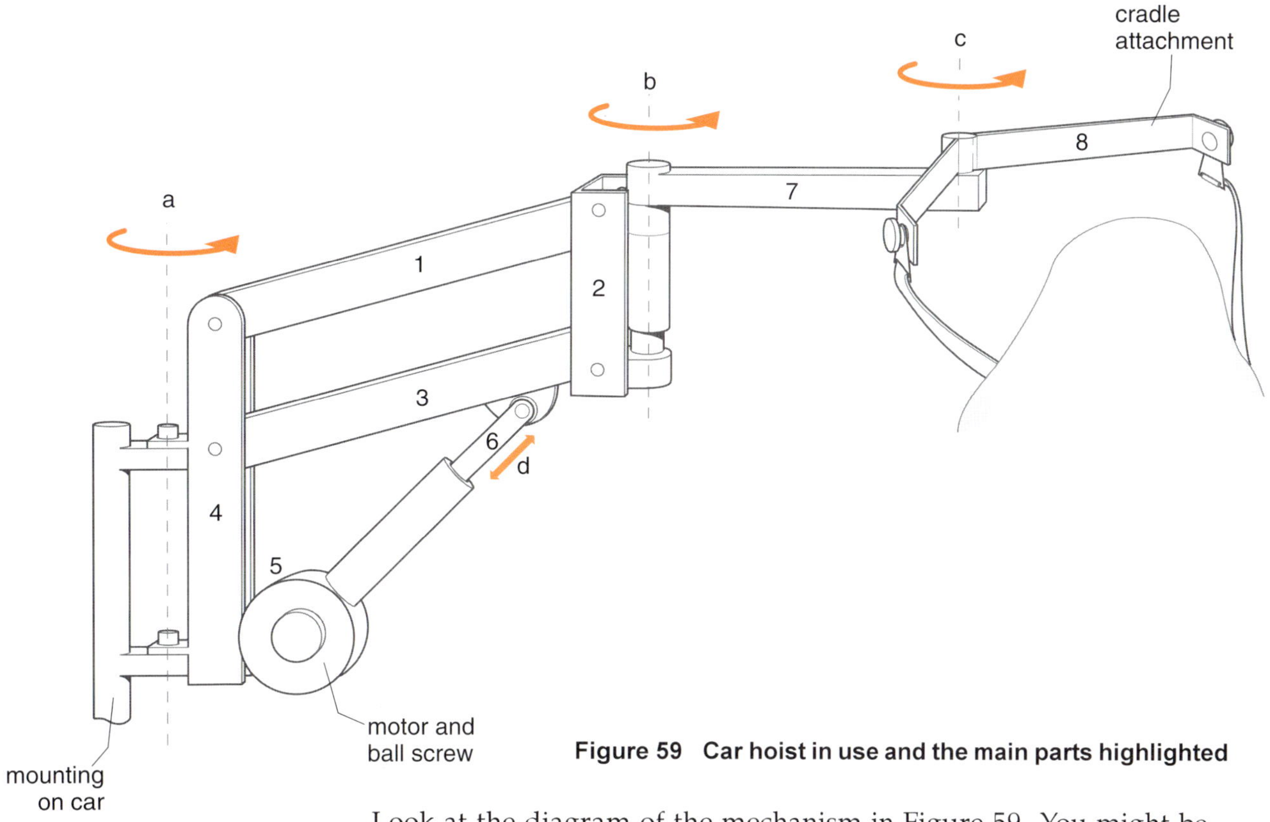

Figure 59 Car hoist in use and the main parts highlighted

Look at the diagram of the mechanism in Figure 59. You might be surprised it uses a triangle that, given the earlier discussion of rigid structures, you might think is not suitable for moving a load. But you can get controlled motion from triangles if you let the length of one of the sides of the triangle change. In the hoist this is done by an electric motor and a ball screw that changes rotational motion of the motor into a linear motion along the direction marked 'd'. The linear motion lengthens or shortens one side of the triangle. The details of how the ball screw works are not important here but in simple terms it is similar to turning a nut on a bolt, rotating the nut moves it up or down the bolt. The top part of the hoist is a parallelogram. The whole hoist can rotate about the axis marked 'a' where it is mounted on the door pillar of a car. The cradle can rotate about the axis marked 'c'.

Exercise 16 Analysing a hoist

Draw the network of relations between the main parts in the car hoist. For the network diagram treat the connections between moving parts in the same way as you treated connections previously.

Discussion

Although the ball screw assembly has several parts, it has two main parts – the body labelled '5' in Figure 59 and the link (6). The part of the hoist attached to the car (the mounting labelled '4') has three connections, the upper parallel arm of the hoist (1), the body of the ball screw (5), and the lower parallel arm of the hoist (3). See my network diagram in Figure 60.

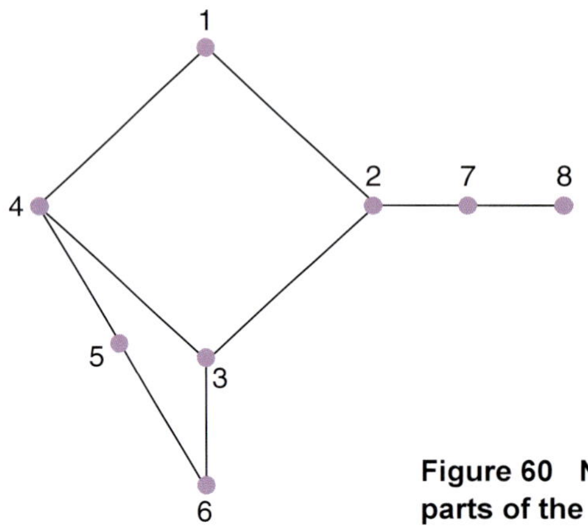

Figure 60 Network of relations between main parts of the car hoist

W 2

This would be a good point to read Sections 3 and 4 of the *Shapes and Space Workbook*. These are titled 'How elements can come together' and 'Making one element move another'.

You may find it helpful to watch the video sequence on James Dyson's design process, 'James Dyson: designer and innovator'. In particular you should pay attention to where he explores possible configurations for the handles, tools and hose connections. He makes extensive use of physical models and prototypes to help this exploration. Notice in this sequence one idea leads naturally to another.

2.5 Possible configurations

Designers use sketches, drawings, physical models and prototypes to help explore possibilities. In some cases, such as James Dyson's accounts of his design process, the generation of one idea leads naturally to another, mainly by trying things out with physical models. The final design emerges from this process. However, it is also possible to explore possible configurations more systematically, especially for simple designs with few parts. I'm going to use house plans again to illustrate this.

It can be relatively easy to generate the possible configurations of rooms in a house plan because of the many practical constraints. Suppose you wanted five rectangular rooms within a rectangular house boundary.

Figure 61(a) has two possible plans with rectangular rooms and constraints as follows. The entrance hall (1) must face the front and have access to the sitting room (2), the dining room (3), and the kitchen (4). Access to the utility room (5) will be from the kitchen. Each room should be next to an outside wall of the house. These configurations are not dimensioned and just show the arrangement of rooms.

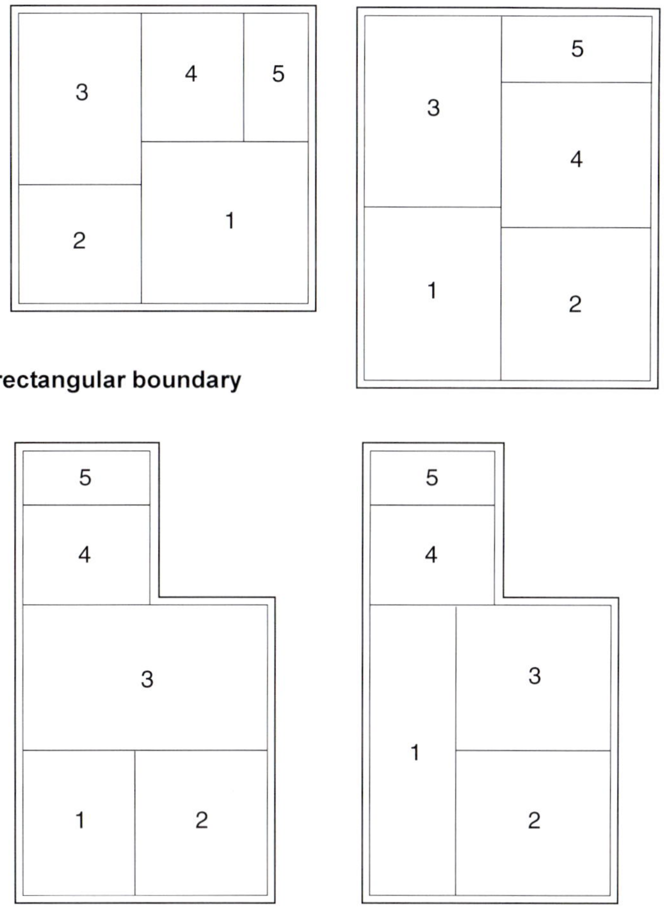

(a) Within a rectangular boundary

(b) Within an L-shape boundary

Figure 61 House plans with five rooms

Suppose you wanted five rectangular rooms within an L-shape house plan. The five-room plans in Figure 61(b) of terrace houses are within an L-shape boundary. These terrace houses have narrow frontages and there can be only two rooms facing the front. Further, the entrance hall faces the front and has access to rooms 2, 3 and 4. With these constraints few configurations are possible (two are shown in Figure 61b).

2.5.1 Three-dimensional configurations

The configurations you have seen so far are essentially two-dimensional, but most products have three-dimensional (3D) configurations. For example, a house often has two floors. The plans of the two floors are not independent. Walls are often aligned on the two floors and the stairs form a vertical connection between floors.

Configurations that are even more complicated are the arrangements of engine components under the bonnet of a car. However, most 3D configurations have some organising principle. The car engine has an axis along the output drive shaft and another axis in the direction of the cylinders. Similarly a gearbox has several parallel axes for the gears.

Figure 62 Organising principles – layers, axes and symmetry

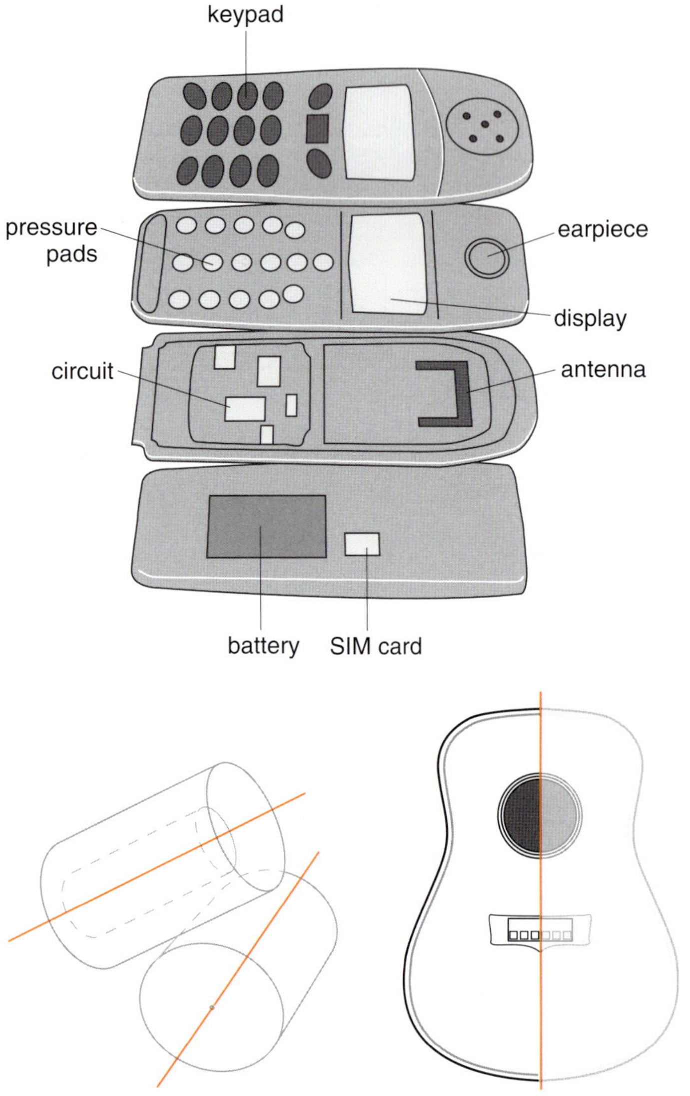

The Dyson vacuum cleaner has a configuration of interpenetrating cylinders, and the Strida bicycle has a triangular configuration which folds in three dimensions. In designing these 3D configurations, visualisations in the form of physical models or computer-aided design (CAD) models are invaluable.

2.5.2 Organising principles

A designer might use an organising principle to help create a 3D configuration. Figure 62 shows examples of three organising principles:

- layering – as in the main parts of a mobile phone;
- arrangement of axes – for example the two axes in the Dyson vacuum cleaner;
- symmetry – for example a guitar or a pair of semi-detached houses.

These principles are not just geometrical but reflect how the product works and how the product will be used. For example, separate floors in a house reflect how that building is expected to be used. When someone wants to use a house in a different way this may give rise to an open-plan or galleried house design.

2.5.3 Cataloguing all configurations

Sometimes it is possible to generate all possible ways of configuring parts. Figures 54 and 63 compare two catalogues of possible configurations. Figure 54 is the catalogue of possible configurations in the patent application for the Strida bicycle. Figure 63 was created by Felix Wankel for rotary piston engines. It is just one section of a larger catalogue of possible configurations.

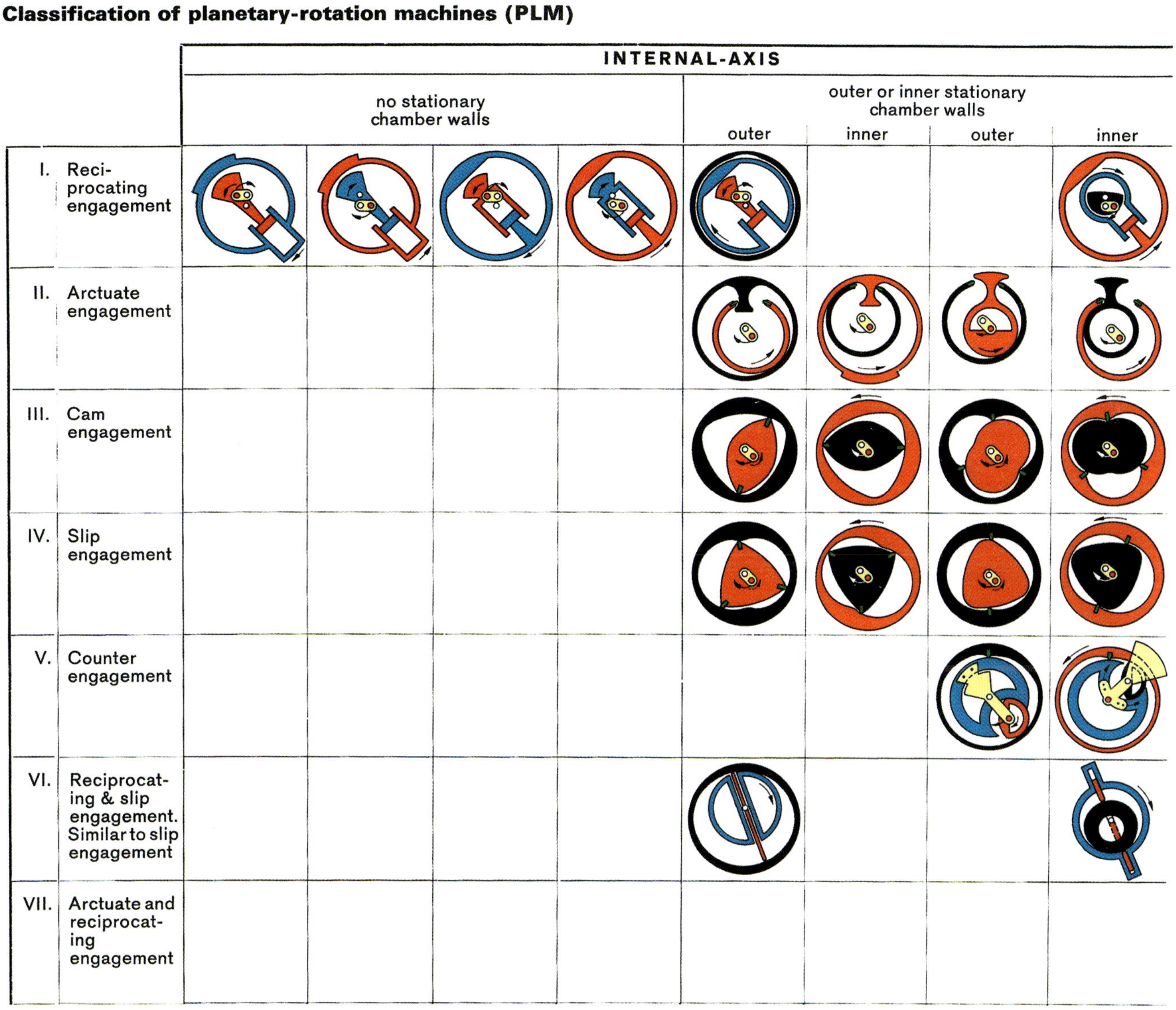

Figure 63 Wankel catalogue

There is an interesting difference between these two examples. The Wankel catalogue was drawn up to summarise and record a long and detailed search by Felix Wankel for successful rotary piston machine designs. Its purpose was to record the results of design, experimentation and testing in the table of possible configurations. The Strida catalogue was generated systematically in the process of design to ensure that possibilities, which might be successful, were not overlooked. It was also to protect ideas from being copied.

2.6 Different layouts with the same configuration

After a configuration has been generated there may be a need to vary the shape but retain the configuration. Changing the size of each room in a house plan, one by one, is tricky. Changing one room will cause changes to other rooms while the configuration remains the same.

Figure 64 shows one way to change the sizes of rooms but preserve the configuration. The walls will lie on a grid, albeit a rather irregular one like a tartan. Changing the spacings between the grid lines alters the sizes of rooms while maintaining adjacencies between rooms and therefore configuration.

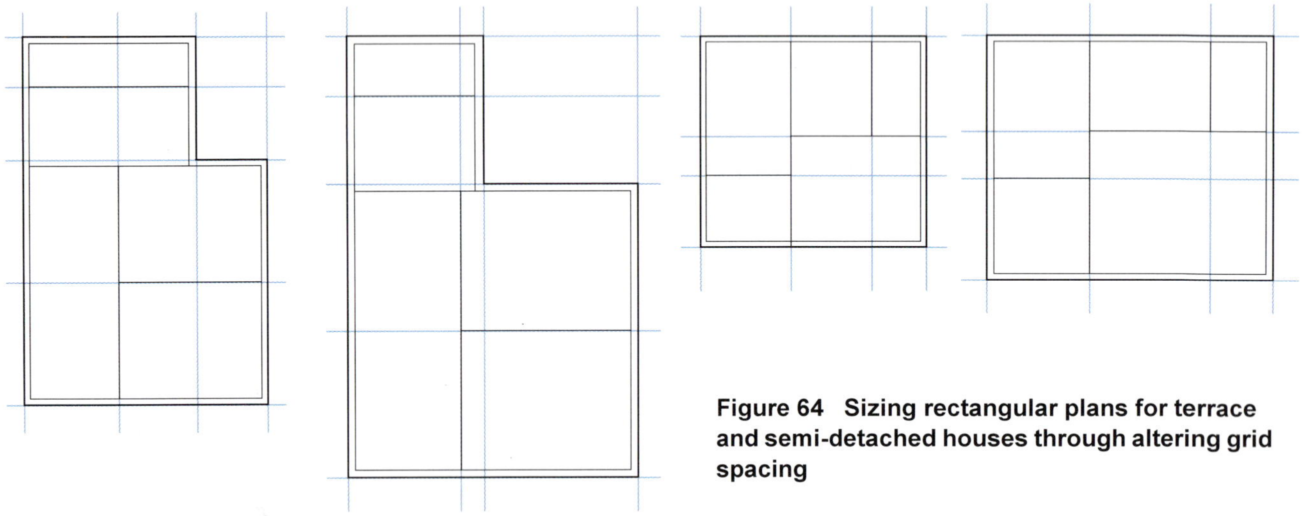

Figure 64 Sizing rectangular plans for terrace and semi-detached houses through altering grid spacing

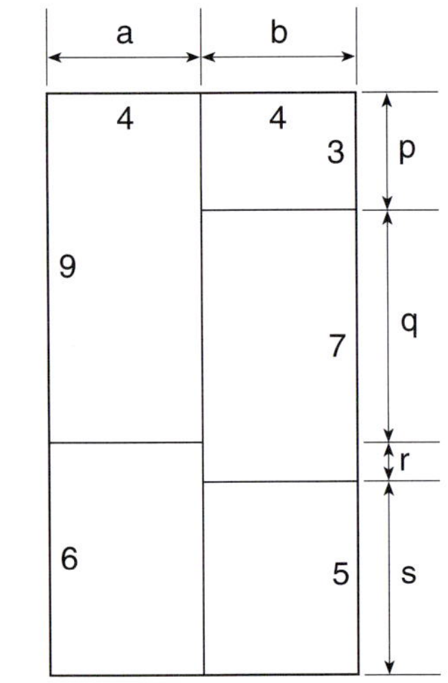

Figure 65 Configuration and sized plan

SAQ 6

Figure 65 shows room dimensions. Around the perimeter are letters standing for the distances between walls. Work out the distances a, b, p, q, r, s from the dimensions of the rooms shown on the plan. For example, a = 4 and p = 3.

It is possible to do more than alter the spacings between the lines of a grid. The straight grid lines can be changed into curves. This distorts the drawing. These metamorphic transformations are helpful in generating versions of the same configuration. The metamorphing of shapes is a well-established technique, as you can see from the drawings of fish by D'Arcy Thompson from 1917 (Figure 66).

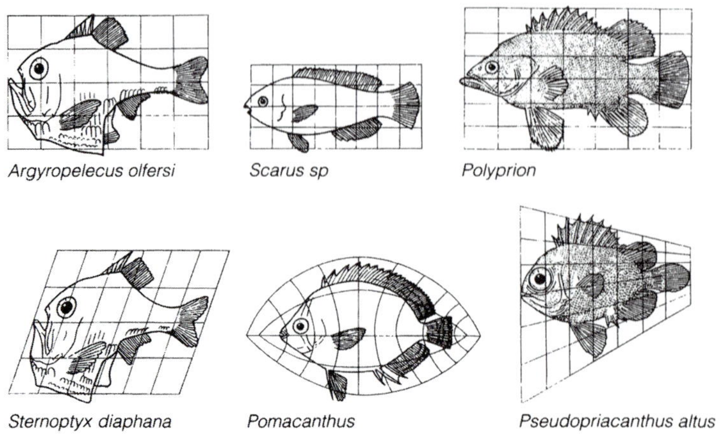

Figure 66 Varying a grid over drawings. Fish drawn by D'Arcy Thompson

Suppose I wanted to explore different layouts by varying the sizes of the parts of the hoist shown in Figure 59. I might be looking for new designs that are similar to the existing hoist and have the same configuration of parts. Of course it is critical to maintain the essential functions of lifting capacity, reach and manoeuvrability. Figure 67 illustrates schematically two variants of the hoist that have some of the lengths of the different parts altered while the reach was kept the same.

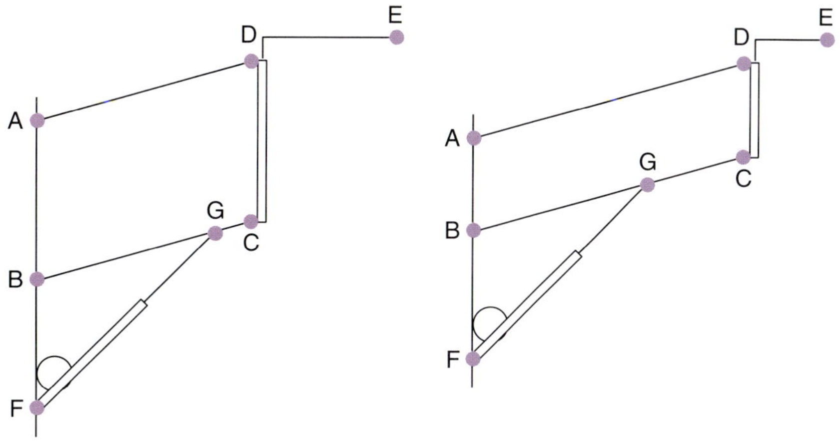

Figure 67 Two variants with different dimensions of the electric person hoist shown in Figure 59

Notice the variant shown that has the longer parallelogram (ABCD) may not be suitable for the reason that the arm DE is not long enough to swing the person without clashing with the hoist itself. Notice also that some dimensions remain the same in the two variants. The mounting in the car along BF is constrained to be the same length in each variant. This would be a sensible standardisation across designs, allowing interchangeability. The motor drive and ball screw (along FG) are also the same in each variant. Again this is a desirable constraint to simplify ordering from the supplier of drives and maintenance of the hoists in service.

In other respects there is freedom to explore different sizes – and possibly shapes – of parts within a configuration for the hoist. For example, a designer might consider different cross sections for the main structural lifting parts, or different connections between moving parts. But the design must continue to meet the essential functions of lifting capacity, reach and manoeuvrability.

Key points of Section 2

Thinking visually and spatially is a key design skill. The exercises on parts, relations and repetition develop this skill.

Different relations between parts give different configurations.

Network diagrams are used to represent configurations.

Relations between parts give rise to rules for generating possibilities.

Repeating parts in systematic ways yields regular patterns.

The basic ways of repeating a motif (a part) in two-dimensions are by rotation about a point, reflection in a mirror line or translation along a line. These can be combined to generate many different repeating patterns.

Three organising principles of three-dimensional configuration have been presented: layering; arrangement of axes; symmetry.

Moving configurations have connections between parts that allow movement. The configuration stays the same during motion – as you saw on the hoist for example.

Moving configurations can be represented by networks of connections in the same way as 'static' configurations that do not move.

A single configuration can be realised by parts with varied shapes and sizes, as long as constraints are obeyed and essential functions met.

Learning outcomes: 1.2, 2.1, 2.2, 3.1, 3.3 and 4.1

W 2

This would be a good point to read Section 5 of the *Shapes and Space Workbook*, 'Topology, morphing and proportions'

3 Function

In this section I have used few exercises and there are no references to the workbooks or supplementary materials. This will allow you to read through the material without breaking for other activities.

A configuration of parts can explain a lot about how a design functions. In my use of the word function, I will make it work for me in a broader way than the author of Block 3 did.

Function is more than what a product does because it is also about how it is used and who is using it. A terrace house configuration with a narrow hall and passage from front to back may be used differently from a semi-detached house. However, there is far more to how a house functions than is revealed by the configuration alone. The sizes and proportions of rooms, materials and energy consumption are all properties of the house that determine how it functions.

You saw the same in the configuration of the car hoist in Section 2. Configuration revealed the two main mechanical functions: lifting by a driven, controlled motion; swinging by manual means. However, when you look at the details of the functions a user of the hoist requires then shapes, sizes, materials and other properties become important. What are the detailed functions that a user requires in a product? I will look at some other products.

The function of a fashion garment might be for comfort or to make an impression. The way the parts of a garment are configured directly influences these functions. However, the precise effect of a garment is then determined by colour, fabric (material) and the shape of parts such as sleeves and pockets. These are functions of appearance and feel – how a design is perceived and used.

Appearance of a garment comes from several sources. I mention just two. First, there are technical properties of fabrics, such as optical effects created by the weave. For example, the way that a satin changes its appearance under different lighting conditions and when viewed from different angles is a direct consequence of how the fabric is woven. Second, the way the fabric is cut (shapes and sizes) and joined (connections) determines how the garment appears when worn.

For a household product to function well it has to meet user needs. The vacuum cleaner has parts such as motors, filters, cyclones and dust bags that are configured to provide the technical function of sucking dirt from the floor, furniture and curtains. However, this technical function and the associated configurations of parts are pretty useless without attention to how people actually use a vacuum cleaner. The configuration of the controls is designed to allow easy use, with a foot switch perhaps. The connection and disconnection of attachments must be straightforward.

A vacuum cleaner is used frequently and the user wants an object that appears attractive and is comfortable to use. The manufacturer also intends that a product has its distinctive brand appearance.

3.1 Types of function

3.1.1 Distinction between functions and properties

Some requirements are derived from how a product should function while others are descriptions of properties of a product. For example, a particular size of engine in a car and legroom in the back seats are product properties. However, they are pretty close to being functions. Corresponding functions would be that the car performed with a certain acceleration, or could carry back seat passengers in comfort over long journeys.

Notice the distinction between property and function. The first describes some actual property of the car – the size of its engine or legroom – without reference to the reason why this property was required. The second gives the reason but does not specify what particular property is used – a car could have a certain acceleration because of its weight rather than its engine size. You came across the distinction between properties and functions in Block 2, with the writing of a brief.

In designing, properties and functions are both vital but they need to be distinguished from each other. In Block 2 you saw the sewing machine designed by Kenneth Grange. One of the stated requirements of the new design was to provide support for fabric in front of the needle. One of the properties (in this case a shape property) of the machine is the asymmetric geometry that emerged from the design process. You can see how the function (support for fabric) and shape property (asymmetric geometry) must be considered together during designing.

Functions are delivered by the properties of the product. I will return to how properties of components deliver functions later and here I'll explore how to analyse the functions of a design.

3.1.2 Essential and secondary functions

Some functions are essential. A kettle boils water, a vacuum cleaner cleans floors. These are the essential functions in the sense used in Block 3. Some functions are secondary such as a washing machine might have a range of functions incorporated in the settings to wash different types of fabrics. This introduces yet other functions. To deliver the essential function of washing clothes necessitates other secondary functions such as timing wash cycles and controlling temperature.

This hierarchical sequence of functions can be unpicked even further for the washing machine. The timers are either electronic or mechanical. If mechanical, they often consist of rotating cams that operate switches. Timing requirements lead to further details on the placement of cams and switches. The next level of detail is how the cams and switches work. So as well as the essential function for a product, there are the secondary functions of many parts.

3.1.3 Five types of function

There are five types of function for a washing machine.

Technical

Technical functions are expressed in the language of engineering – motion, force and power, strength and so on.

Usability

Usability functions deal with how the product is used. These functions include, for example washing programmes, layout of controls and ease of loading.

Sustainability

Sustainability functions deal with the environmental impacts of energy and materials consumed during the life of the product. The wider implications of sustainability will be explored in Block 5.

Aesthetics

Aesthetic functions deal with how people respond to appearance, feel and even sound. For a washing machine this will be mainly about the user interface – the front of the machine, the controls, the door and perhaps the inner drum the user comes into contact with when filling and emptying the machine. Colours, shapes and textures are especially important to aesthetic function.

Identity

Identity functions deal with the product's identity. One aspect of this function is recognising a brand, with all its associations. In products such as cars the creation of a strong identity is important for both the individual car and a range. With many products, washing machines for example, brand recognition, identity and personality play an important role in consumer purchasing decisions.

3.1.4 Three applications for functions

The five functions of technical, usability, sustainability, aesthetics and identity can be applied in three ways depending on what the designer is concentrating on.

Application 1 – a future product

In this application the designer identifies which functions need the closest attention because the specification drew attention to them. Paying attention to usability, say, would raise the question, what are the user needs?

Application 2 – an existing product or prototype

In this application the designer analyses how the product functions in use. Paying attention to identity, say, would raise the question of what is the brand status. Paying attention to technical functions would raise a question like what is the maximum speed or load capacity of a design.

Application 3 – a product being designed

In this application the designer identifies which functions to consider during the design process and predicts how a design proposal might function. Paying attention to sustainability functions, say, would raise the question, now that I know I need that part there, what type of plastic can I use for the part so that it can be recycled?

3.2 Technical function

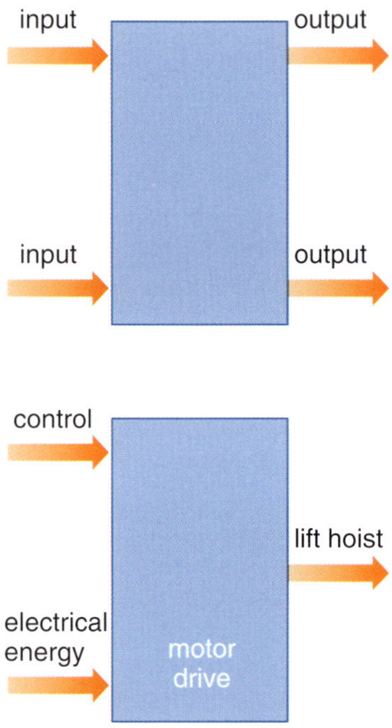

Figure 68 Technical function relating input and output

To give a better idea of how a designer deals with functions on the way to arriving at a layout, I will say some more about the five functions, starting with the technical function.

Technical functions concern how a product works. Technical function can be pictured in terms of inputs and outputs, for example by a nontechnical diagram like Figure 68.

There is no shortage of examples of technical function. A car gearbox takes the action of the car engine as input and transforms it to output as a motion at the right speed to drive the wheels. The person hoist (Figure 59) contains a motor and drive that takes electrical power from the car battery as input. The output is the extending motion to lift the arm of the hoist (Figure 68).

Technical functions are achieved by an electronic mobile phone when key presses as inputs are converted to output signals that are transmitted to a network. Another technical function of the phone is in the other direction, when an input signal from the network causes the phone to output the audible ring tone.

Some technical functions involve individual components. A component, for example a plastic part for a children's toy, may have the function to be strong and tough in the sense that it should not break if trodden on and even if it does break should not fracture dangerously into sharp pieces. Again you can see this in terms of an input and an output: a force on the component; a response to the force by bending and temporarily deforming but not fracturing.

Technical functions are important in design. Motors, gearboxes, electronic circuits and materials can be designed to give a range of different functions. There are large amounts of available data to help designers meet these technical functions. One difficulty facing designers is finding this data for a particular design. Another is using the available data effectively given that a design not only needs to function technically but also with regard to usability, sustainability, aesthetics, and identity. With large resources to solve the problems of technical function there may be a temptation to concentrate on this aspect of function.

Generally technical function can be specified precisely. For example, gear ratios, power output at different speeds for a motor, characteristics of an electronic circuit or material properties all have exact meanings and can be described quantitatively. Likewise, once a design is completed, technical functions can be assessed. For example, functional performance of a gearbox under different load conditions can be evaluated largely from computer models and simulations and then confirmed by testing. For instance with a portable electric drill the power output at different speeds can be predicted accurately. Heat to be dissipated, strength of the casing, recharging time and battery life can all be evaluated before a design goes into production. But these latter functions are not just technical, they are also user functions. Recharge time and battery life will affect usability. So the different types of function affect each other.

3.3 Usability

To illustrate this function I will look at a finished product and examine how the configuration and the properties of parts affect how a product is used.

To be held in one hand and be moved around easily, a portable electric drill with a battery should weigh less than 2 kilograms. In the example shown in Figure 30 the handle allows a firm handgrip, controls are part of the handle, and the battery is positioned so that the whole drill is balanced. These user functions are about the properties concerned with size, weight and shape – the properties adopted should make the drill easy and convenient to use.

3.3.1 Functions of a user interface

The way products are used depends on how the user controls the product. The interface for controlling a video cassette recorder is an example. Such an interface is a major design in its own right.

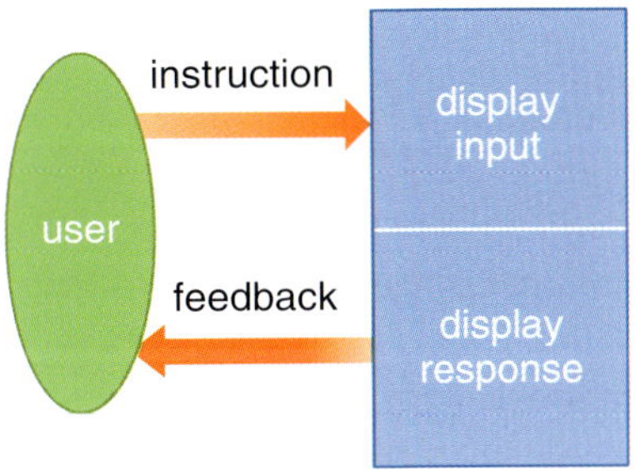

Figure 69 Three elements of a user interface

What are the main elements of user interfaces? First, there is a means for the user to give instructions. This might be by pressing buttons, setting values or pointing to items on a display. Second, there is a means to show instructions have been received by the device. Third, there is a means to give feedback on how the device has responded to the instructions. Of course the feedback may show an error and the user can modify the instructions and try again. This is shown with the simple diagram in Figure 69.

Sometimes I don't only use a product's visual display. When tuning a radio I use a display to get near the station I want and then fine tune by listening for best reception. This twiddle-and-listen method can now be done automatically by electronics that search for the best signal. Typically a car radio is tuned this way because the driver should concentrate on driving.

The creation of inclusive designs that are accessible to people with a range of physical limitations – sight, hearing, manual dexterity, mobility, and so on – requires careful thought to the design of user interfaces. The size, shape and arrangement of knobs and buttons; the colours used in controls and displays; the presence of visual, audible and tactile responses, are all critically important for usability.

3.3.2 Side effects

There is another important aspect that I alerted you to earlier. When using a product there are often side effects that make it unpleasant or even dangerous. For example, a vacuum cleaner might make a lot of noise, an electric drill might get rather hot to hold, a tablet computer or laptop computer might become uncomfortably warm if rested on your lap for a long time, shiny kitchen appliances may reflect glare from spotlights, a nice texture to grip for a kitchen utensil may make it difficult to keep clean.

These side effects are often the result of decisions on the detail of designs. Accordingly they can be fixed by attention to details. Noise in your vacuum cleaner may be from the motor or fan, or from vibration of the casing. Detailed changes to these parts may reduce the noise. However, if changes to the source of noise are not possible, a designer might seek to reduce its effect on users by inserting a sound absorbent material on the inner surface of the case. But fixing side effects like this is not straightforward – fixing one side effect can introduce others.

As I've mentioned noise I will digress on the subject a moment. For some products the type of noise made has become a desired property. Partly noise can reassure users of the condition of the product. I'm sure you can think of one of your own products such as a vacuum cleaner or other motorised product that has caught your attention because it didn't sound normal. Also, in Block 2 you came across cars that had their exhausts tuned to a sound that was intended to appeal to a certain segment of the market. Sound can be a subtle but powerful influence on user perceptions. Controlling and predicting sound is a specialised science and some companies will devote considerable attention to testing sound in their products.

3.3.3 What noise annoys?

For any stimulus there is usually a threshold that causes physical pain. Because of the relation of stimulus and response this threshold is often many millions of times larger than the smallest stimulus that can be heard or seen. Where the range of the stimulus is so great, from small values of stimulus to large values, special scales are used for measurement. For noise stimulus the special scale is called decibels (dB), and goes from 0 dB (threshold of hearing) to 120 dB (unbearable noise, causing severe pain).

The way this works is that when a measuring device shows an increase of 10 dB, the actual stimulus has gone up 10 times. A quiet whisper is about 30 dB and normal conversation is about 60 dB, so the sound stimulus reaching the eardrums in the conversation is 1000 times (10x10x10) more than the stimulus in the whisper. Each 10 dB increase means the stimulus is multiplied by a factor of 10. A domestic vacuum cleaner is between 70 and 80 decibels. That means it is between 10 and 100 times the stimulus of a normal conversation. A lorry at 10 metres gives a noise of about 100 dB so this is 10 000 times (10x10x10x10) more stimulus than a normal conversation.

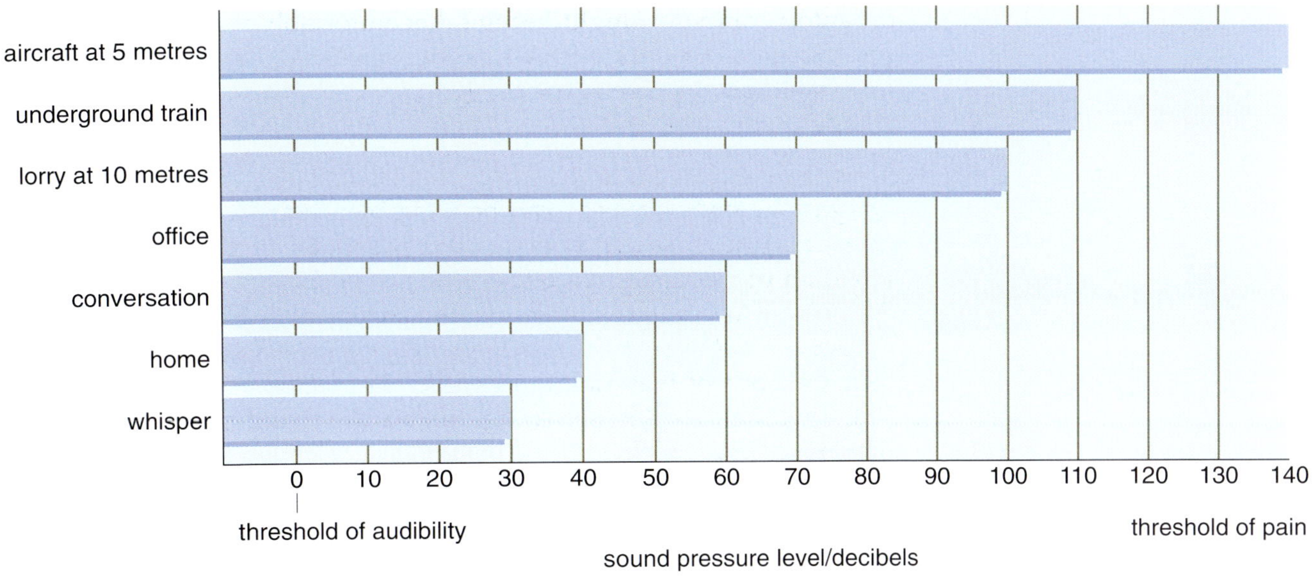

Figure 70 Decibel scale

3.4 Sustainability

As you saw in Block 2 the life cycle of a product refers to all the stages of material extraction and processing, manufacture and distribution, use and maintenance, and disposal and recycling. This section concentrates on just one aspect of sustainability – the environmental impacts of a product during its life cycle. Many of the impacts are the direct result of decisions at the design stage but estimating actual environmental impacts during design is difficult and results are only approximate.

3.4.1 Clean clothes at what cost?

Environmental impact differs from other types of function because it affects many people, not just the user. Consider again the washing machine. It consumes energy, water and detergents through its life. Producing the energy is likely to create pollution and carbon dioxide emissions without immediate impact on the user. Water is a natural resource that is purified for use in treatment works, which consume energy. The detergent is manufactured in a process that is likely to create some waste products that require disposal. The detergent comes in packaging that also needs disposal. The dirty water will require treatment to remove dirt and detergent before it can be used again. The dirt and detergent requires disposal. All of these are environmental impacts from using the washing machine.

There are further impacts from the resources and energy used to manufacture the machine and its disposal at the end of its life. It would be desirable for most of the materials in the machine to be recycled. However, although possible in principle, this may not be done unless regulations and legislation make it mandatory.

Environmental impacts are complex and have many facets. Energy use can be directly costed in terms of prices of the electricity, gas or oil required. This is a direct cost to the person paying the fuel bill after the washing machine has consumed the energy. However, more general environmental costs will be needed to measure wider impacts.

When I use my washing machine I pay for electricity, water and sewerage but this may not completely reflect the long-term cost of associated pollution, reduced quality of life or harm to health. These environmental costs are difficult to estimate and the way of measuring costs of greenhouse gas emissions will be different from measuring costs of toxic pollutants.

3.4.2 Estimating environmental impact

Methods and tools have been devised to estimate environmental impacts. As part of the DVD material for Block 3 you came across the MET matrix, which can assist designers to explore potential environmental impacts. Another is the eco-indicator, which is a multipurpose tool that can be used when evaluating a design proposal, when making design comparisons between one design and another or in identifying parts of a product that are generating high environmental costs. You will be introduced to the eco-indicator software on the DVD in Block 5.

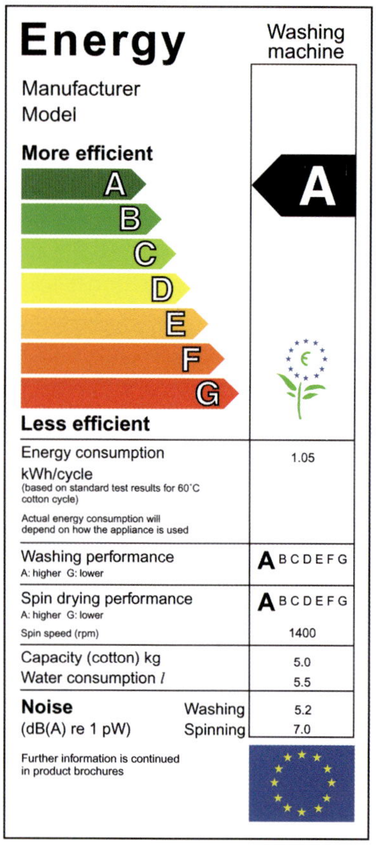

Figure 71 Energy label for a washing machine

I will examine here one environmental impact, namely, energy consumption in using a product. My washing machine was evaluated for energy usage during its design and when I bought it there was an energy rating label stuck to it. Regulations require that each model is tested and labelled in a standard way to indicate energy use (Figure 71 is a label for an appliance that has the 'more efficient' rating of A). You may have noticed the labels on appliances in stores – catalogues and consumer reports on products also show energy rating.

What energy does a washing machine use? First, there is the energy needed to rotate the drum and pump the water around the machine. Second, there is the energy needed to heat up the water. Energy consumptions for different washing cycles are easily calculated. They depend on temperature of the wash, temperature of the incoming water, amount of water used, and the time spent washing.

There are several other environmental impacts I touched on above, including water treatment. The key to assessing overall environmental impact is to be able to add together different types of impact, for example energy and waste treatment. Indeed, this is the main purpose of the eco-indicator methods, which provide equivalences between different impacts. These problems do not arise here as I am concentrating on just one impact, namely the direct energy consumption. Block 5 will consider how a range of impacts for a product can be added to give an overall environmental impact.

The labels for a washing machine focus on just one cycle – the cotton cycle at 60 degrees Celsius. A typical machine will consume about 1 kilowatt hour per cycle. At one cycle per day over 10 years, this is about 3000 kilowatt hours. This costs approximately £200 at current prices in the UK. I use about £0.07 per kilowatt hour as a representative cost. These costs do not reflect true environmental impacts. With many other environmental impacts during operation, such as water use and treatment, the major environmental impacts of the washing machine come from its use. Designing to reduce energy consumption would help reduce environmental impacts.

3.4.3 Legislation means rethinking vehicle design

European legislation is affecting those designing electrical appliances. There is pressure to consider environmental impact during design because of directives on recycling, emissions and pollution. One such piece of environmental legislation in the UK derives from the end-of-life vehicle (ELV) directive from the European Commission. The following is taken from the directive.

Directive 2000/53/EC of the European parliament and of the council of 18 September 2000 on end-of-life vehicles

Article 7

'Reuse and recovery', paragraph 2

2 Member states shall take the necessary measures to ensure that the following targets are attained by economic operators:

(a) no later than 1 January 2006, for all end-of life vehicles, the reuse and recovery shall be increased to a minimum of 85% by an average weight per vehicle and year. Within the same time limit the reuse and recycling shall be increased to a minimum of 80% by an average weight per vehicle and year;

for vehicles produced before 1 January 1980, member states may lay down lower targets, but not lower than 75% for reuse and recovery and not lower than 70% for reuse and recycling. Member states making use of this subparagraph shall inform the commission and the other member states of the reasons therefore;

(b) no later than 1 January 2015, for all end-of-life vehicles, the reuse and recovery shall be increased to a minimum of 95% by an average weight per vehicle and year. Within the same time limit, the reuse and recycling shall be increased to a minimum of 85% by an average weight per vehicle and year.

To meet these targets for reuse and recycling it is essential the different parts of a vehicle can be disassembled and different materials separated. If two parts made from different materials are joined rigidly together with rivets or adhesive, costs of disassembly may be large. A design to meet the conditions of the ELV directive would attempt to use the same material for the two parts or even merge the two parts into a single, continuous part.

Reuse and recycling are concerned with the end of a product's life. At the conception of a product a designer might look for materials that are renewable as well as being recyclable. Natural materials that can be grown are renewable. These may be preferable to extracted raw materials. Reused materials are also an option – perhaps waste from other products.

3.5 Aesthetics

The aesthetic functions make a product beautiful, stylish or pleasurable to use. You might view this as part of the usability function. However, the aesthetic function represents something about a design that is more than useful. Think for a moment; perhaps you have a product that has an attractive colour that does not add anything to its usability. Alternatively a simple and elegant design for meeting a complex set of technical, user and sustainability requirements may be thought 'beautiful' for reasons other than looks. I will concentrate here on the appearance of a product, with some discussion on the feel of a product's surface.

The appearance of a product is determined by the shape, colour and surface texture. Individual parts have their own aesthetic functions but so does the whole product through the ways the parts are configured. Putting together parts of contrasting and complementary shapes, colours and textures gives the distinctive appearance of the whole product.

3.5.1 Proportion and curves

Proportions are critical to the overall appearance of a product. Take a look at the shape of the book you are reading. It has particular proportions. It is longer than it is wide. Decisions made about this shape are partly concerned with the ways that you use the course material as a student, but also about the appearance of the page as you open the book and read it. This appearance should help you read the material but there are many possible proportions for books.

Three-dimensional, curvy shapes are more complicated. Look at the surface of a common product, such as a television, tool or kitchen appliance. You will probably see many different curves on the surface form. Take the television – the corners may be rounded and the outline may have gently curving edges. Light and shade will emphasise these features to create the product you see. Like a book, the TV screen will have particular proportions. Again, some of these shapes are directly concerned with use while others are about trying to make the television look attractive as an object in your room.

3.5.2 Texture and colour

The surface texture of products also affects their appearance. Look again at one or two plastic products. Few plastic surfaces are entirely smooth. Such products exploit texture and this can range from rough to high gloss. Manufacturers use texture to hide moulding imperfections or to distinguish one part from another as well as for aesthetic reasons.

As with texture, manufacturers have a range of colour options available to them, but for many consumer products the application of colour stays within established parameters. For example, most computers are available in off-white, grey or silver. It's not that companies cannot make multicoloured products. In this case computer companies have confined themselves to aesthetics that are tried and tested in the market place. Apple Computer made a bold and successful break with tradition when they launched the iMac, with its smooth, coloured and translucent casing (Figure 72).

Figure 72 Apple iMacs

3.5.3 Stimulated by surfaces

The aesthetic function includes how something feels and handles. These tactile qualities of a product are powerful. Attracting a customer to handle a product is a characteristic ability of many successful designs. These products have an appearance that invites you to touch – they have a tactile appeal. The surface texture of a fabric in a garment determines how it feels to wear. The weight, thickness and flexibility of fabric determine comfort as well as appearance. Clothes must feel right as well as look good. When choosing clothes I often feel the fabric. When choosing a mobile phone I might hold it in my hand to feel the combined effect of material, texture, weight and layout. This is about judging aesthetics alongside other functions.

Partly the process concerns judgements regarding less tangible qualities. Why do you think the dashboards of some new cars include plastic panels that look like wood? Perhaps it's a hint of history – nostalgia in other words. Perhaps a manufacturer is trying to associate new products with traditional values and qualities. Perhaps it's an ironic reference to what a particular model is *not*. Or is it just an abstract formal styling decision to contrast colours, patterns, textures and surface textures. As you have seen in earlier blocks, complex messages are sent by appearance and feel.

Materials are an obvious contributor to appearance and feel. Steel is usually cold and hard to the touch whereas foamy plastics are warm and soft. Rubbery materials are flexible and have give, whereas ceramics are rigid and brittle. You only have to look at the range of fabrics used to make clothes to realise that different materials can give wide differences in appearance and feel.

Designers know that people can bring associations, prejudices even, to a product depending on the material used. These associations may be as much aesthetic as technical. For example, wood has a natural grainy appearance as well as technical attributes of strength and toughness. An association with luxury may be why wood veneer embellishments are used in cars. Stainless steel has associations of strength and long life. Some plastics have been associated with short life, having rather soft surfaces that show wear quickly.

As with the other types of function already described, aesthetic functions are heavily influenced by decisions on detailed aspects of design. The exact shape, colour and proportions can be critical to appearance. Taking colour, small variations can change appearance. You only have to think of choosing clothes or packaged goods on a supermarket shelf to realise that appearance is a strong determiner of customer choice.

Exercise 17 Appearance and feel

Examine a packet of food, such as a box of cereal, a plastic milk container or a can of soft drink. Set down five features of these products that determine their appearance and feel. Comment on how important appearance and feel of these products is to you when buying and using these products. Further, comment on how important you consider the exact colours and shapes used. Imagine the colour was slightly different from normal on the packaging of a product you buy regularly; would you still choose it?

Discussion

I looked at a cereal packet and noticed the features of the shape and proportions of the box; the colouring and graphic design; and the brand identity. I looked at a drinks can and noticed the texture of the shiny, smooth surface of the can and the texture of the top with its ring pull; its diameter for my grip; its feel to touch when full; and the curved base surface.

You might have noticed that colours are precise. Take two of the same cereal boxes and you will notice hardly any difference in colour. Most people are good at noticing small variations in colour. If the colour varied on the shelf from week to week, even by a tiny amount, you would notice and might think twice about your purchase because you might think there would be a similar variation in content. This is about marketing and manufacture of the product but also demonstrates that designers' detailed decisions on appearance in general, and colour in particular, are important.

3.6 Product identity

When you did Exercise 17 I hope you became aware that the appearance and feel of a product are not just abstract properties that give pleasure or make the product easy to use. Appearance and feel distinguish a product from competing products, and they give it product identity.

Product identity is central to design. The properties of parts – their shapes, colours and materials – as well as their configuration are the building blocks of product identity.

Designers strive to provide identity for a product. This can help make the product sell by distinguishing it from similar products with similar technical, usability, sustainability and aesthetic functions.

Identity allows consumers to recognise a product and it facilitates communication – both from supplier to purchaser and between consumers. The former will be needed on a crowded superstore shelf, in a large catalogue or among thousands of product images on websites. For example, cans of soft drink have the same shapes with similar surfaces and textures – clean and inviting. The colours and surface designs pick out one product from another and give the product its identity.

An identity shared across many products is part of a brand. You have seen a good example of product identity in the DVD video sequence from Philips that deals with the animal-shaped web cams. Taking a broader perspective, an identity shared across brands, and perhaps across a culture, can be called a style.

3.6.1 Individuality, brand and style

I've condensed the notion of product identity into these aspects:

- *individuality* of a single product;
- *brand*, which applies across a range of products;
- *style*, which spans products, brands and cultures.

3.6.2 Individuality

The individuality of a product lies in its particular configuration, shapes and colours. Dyson vacuum cleaners have a distinctive identity created by compositions of cylinders, colours and transparent materials. This identity may be shared across several products in a product range and in some respects the Dyson brand is still apparent in widely differing products, such as the company's washing machine. A similar situation occurs in many companies where products share similar characteristics.

3.6.3 Brand

Brand is a characteristic of a range of products from a particular company. The brand makes them recognisable, and perhaps creates associations with reliability, modernity or fun. Manufacturers want to remind you of previous products that have been successful.

A brand might be just a name such as Hewlett Packard or Dell for computers. These don't really have a particular appearance but invoke strong associations for product type, which the companies emphasise in advertising. The computer company Apple does not really have a visual brand – its name is its brand. But it does produce designs with product identity. Other brands have a name as a significant feature of designs. Coca Cola identifies itself by using at least one of the following on each product: the special curly script of the company logo; the shape of its bottles; the distinctive red colour. Other well-known brand elements are shown in Figure 73.

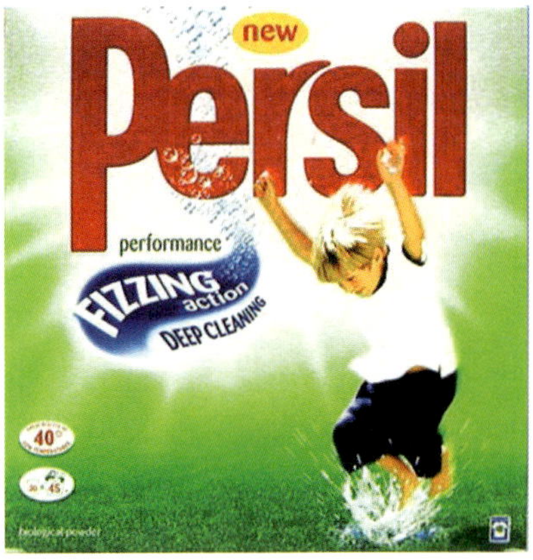

Figure 73 Brand elements: shapes, colours, logos

3.6.4 Style

Style is rather different and goes across products and brands. Style may be characterised by the presence of certain shapes and colours in particular configurations. Styles often belong to certain periods. The art deco style in Europe and north America in the 1920s and 1930s led to common shapes on objects ranging from train engines to refrigerators. In the 1950s the car designs of Harley Earl for the US company General Motors are representative of a style of vehicle design of this period (Figure 74).

Different types of product can all share a style. It can be quite hard to define style precisely in terms of configurations, shapes and colours. Two products from the same style will appear similar. They have similar elements configured in similar ways.

Style is a huge subject. It is more than a convenient way to classify products and designs because, like brand, a style has complex patterns of associations. Lifestyles and worlds of thought are conjured by a style.

Figure 74 Three of Harley Earl's designs from the 1950s

Key points of Section 3

Five types of function are identified:

- technical function;
- usability;
- sustainability;
- aesthetics;
- identity.

These functions of a design can be applied in three ways: as desired functions for a future product; as performance of an existing product or prototype; as predicted performance of a design proposal.

Technical functions can be modelled as relations between inputs and outputs.

Usability is about your interface with products of all types, from controlling video cassette recorders to holding a fork. Interfaces require ways to (a) give instructions, (b) display instructions, and (c) display response of the product.

Environmental impacts, as one aspect of sustainability, depend on: materials used, reused, recycled, recovered; energy consumed during life; waste and harmful by-products of manufacture, use and disposal.

Aesthetics of a product make it beautiful, stylish and a pleasure to use. Among many other factors aesthetics depends on shape, proportions, colours and surface texture.

Product identity is concerned with: the individuality of a single product; brand is across a range of products; style is across products, brands and culture.

Learning outcomes: 1.3, 2.3 and 3.1

4 Component design

Component properties deliver functions. What does this mean? Let me use a familiar example. A component such as the button of a doorbell should be strong enough not to fail in its technical function, which is to allow normal finger pressure to make an electrical connection by pressing the button. Its surface colour and texture may also deliver functions such as assisting visibility and avoiding a finger slipping off. I will say more about colour and texture later.

Many component properties are associated with the material from which the component is made. Take, for example, the strength of the doorbell button. The same button could be made in different materials with different strengths. The property of strength depends critically on material and is referred to as a material property.

Other material properties include hardness, toughness and elasticity. This is only a handful of a large number of material properties that designers might specify for components.

Perhaps the most immediate and obvious property of a component is its shape. This is critical to many functions. Examples of how shape contributes to the different types of function are not hard to find. To perform its technical function a gear needs regularly spaced teeth of a special shape. Usability of a toothbrush depends on the shape of its handle. Aesthetics of a car and its appeal to different customers depend on shape and styling. Product identity of the Coca-Cola bottle comes from shape, colour and lettering. The influence of shape on sustainability is harder to pin down – an example could be an aerodynamically shaped vehicle reducing energy consumption.

4.1 Shape

Some products such as a car have tens of thousands of components: tyres, wheels, springs, steering wheel, windows, seats, safety belts and so on. Other products have tens of components, such as an electric drill, a food mixer, a dress or pair of shoes. Finally there are simple products such as spoons, bowls, or even a tablecloth with just one part. But are these latter products simple because they only have one part? I'll return to the example of a table fork to explore the phenomenon of shape.

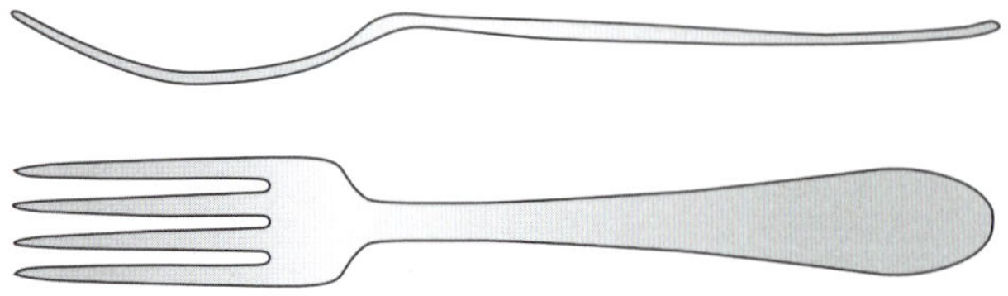

Figure 75 Table fork seen from side and top

Follow the outline of the section of the fork taken through the handle and one prong of the fork (see Figure 75). It curves gently in some places and in others changes direction quite sharply. The outline plan of the fork shows a variety of curves with points at the ends and smooth curves on the handle. And I've only looked at the section and plan. If you actually look at a fork like the one in Figure 75, you see a complicated curved surface that changes from one part to another.

The fork has complex shapes curving this way and that. These shapes are difficult to make by machine – they can require complex and expensive manufacturing processes. I will start with some simple shapes I commonly see as parts of design. These shapes have flat surfaces (not curving at all), cylindrical surfaces (curving at a constant rate just one way) and spherical surfaces (curving at a constant rate in all directions). See Figure 76.

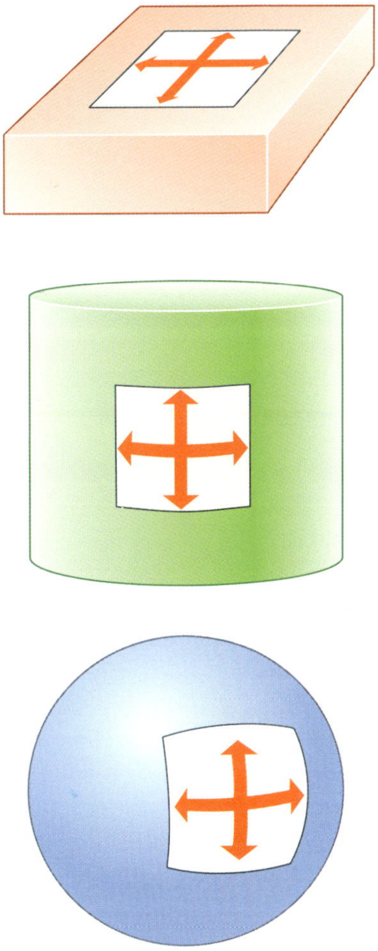

Figure 76 Flat, cylindrical and spherical surfaces

These simple shapes can be made easily on machines and I might call them machine shapes. In contrast, complex shapes suggest growing and living things – I will refer to these as organic shapes.

Complex surfaces are appearing more and more frequently in mass produced products. It is instructive to look at car styling before the use of computers. The Harley Earl studio at General Motors was famous for its car styling in the 1950s especially those outrageous and flamboyant designs with fins and chrome (see Figure 74). Figure 77 shows this same studio slightly earlier. The large blackboard in Harley Earl's studio has a full-size chalk drawing of the car. Curves were drawn bit-by-bit using standard curves. You can see designers holding standard curves and several curves hung up in the foreground.

Figure 77 Harley Earl design studio at General Motors

These standard curves look a bit like odd hockey sticks. They have a gently curving part followed by a sharply curving part. The gently curving part appears to uncoil from the sharply curving part – the whole curve has what a designer calls dynamic tension.

There are two observations I want to make, in case you think I'm out of touch with car design. First, car design is now done differently. Many different models of cars are used at early design stages including clay models, accurate mock ups and a variety of computer representations (Figure 78). The shapes of new cars are different to 1950s cars but they remain complex curving shapes.

Figure 78 Varieties of models used in car design

Second, the lines of the car on the blackboard in Figure 77 are created by the edges of panels in the car. There are well-defined surrounds for the wheels, bonnet and bumpers. These cars were heavy. Today cars are lighter and the steel they are pressed from is thinner. The lines of modern cars are created by details on the panels – raised edges, ridges and corners. These have a dual purpose because they also serve to strengthen the thin panels, giving a lightweight body with less steel than a body with smooth panels.

The rest of this section looks at simple shapes and their surfaces, curvy shapes, dimensions that describe components, the strength of shapes, surface finishes, and colour.

4.2 Simple shapes

This would be a good point to begin the computer-aided design (CAD) activities introduced and described in the *Block 4 Supplement*. You will be directed to solidThinking, the CAD software that is provided on the course DVD. Give yourself a generous amount of time to become familiar with the software. Work through Lesson 1. You'll need to have completed Lesson 2 by the end of Block 4 but you may do this when it's convenient for you.

I will now examine a range of shapes and their surfaces. The first are cylinders and spheres.

4.2.1 Cylinders

Cylinders are common shapes in machines. Cylindrical wheels and shafts are rotating elements that need to be the same all around so that they can roll and turn smoothly. Cylindrical shapes are also easy to make on machines. You will have discovered some of the characteristics of cylinders when you made the cardboard mug in the *Modelling Workbook*.

How would you describe a cylinder? Taking a section across the axis gives a circular disc. A section along the axis is a long rectangle. If the section is taken at an angle to the axis, it is an oval shape (Figure 79).

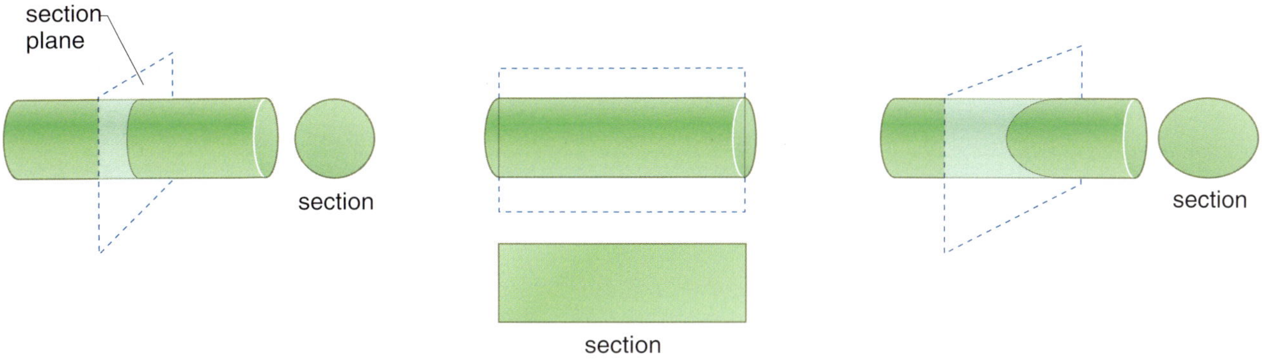

Figure 79 Sections of a cylinder

As the cylinder keeps the same section across the axis along its length, the shape is described by this circular section and the rectangular section along the axis.

Many cylindrical components do not keep the same cross section all the way along. The shaft of the electric motor in a washing machine or vacuum cleaner has electrical windings in its centre, steps for bearings on either side and a mounting for a pulley wheel at one end, therefore the diameter of the cylinder varies (see Figure 80).

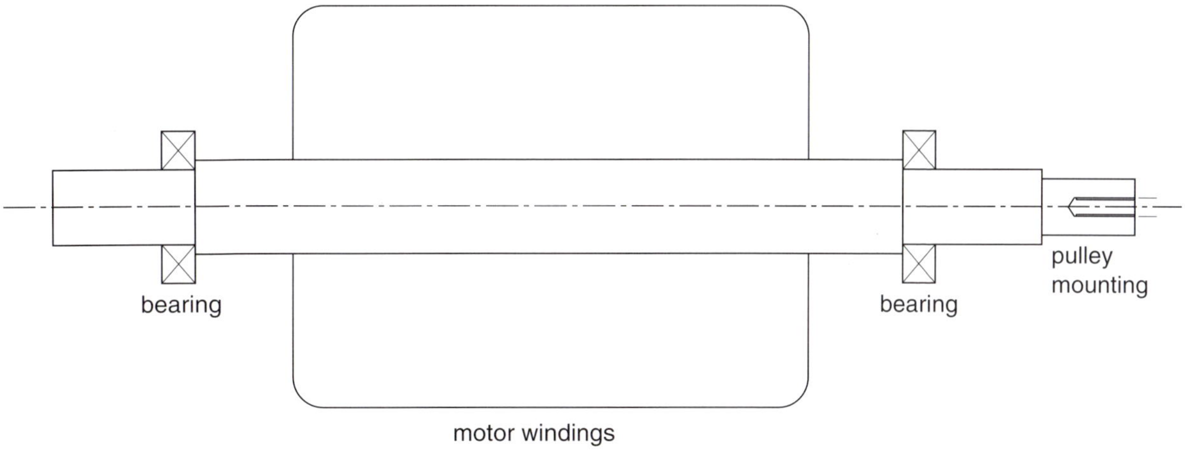

Figure 80 Shaft from electric motor with bearings, motor windings and pulley mounting

Figure 81 Section along axis of soft drinks can

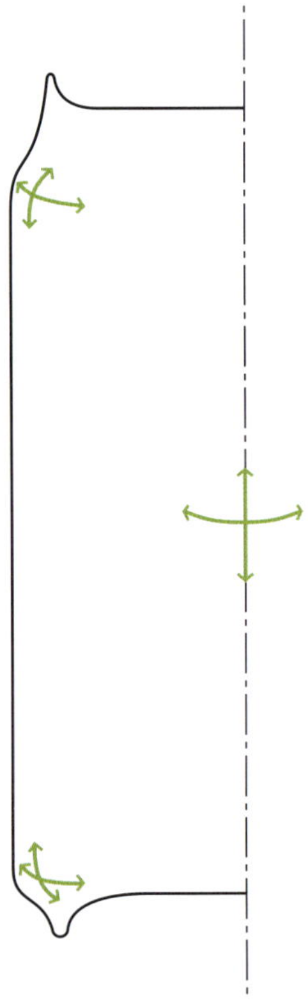

Figure 82 Middle of can has a curve around but no vertical curve; but shoulder and base curve around the can and have a vertical curve

But you don't have to go inside a machine to see complex cylindrical shapes. The *Modelling Workbook* (Section 7) will have drawn your attention to an aluminium soft drinks can with its circular cross sections. These circles change size at top and bottom of the can (see Figure 81). The top of the can is a separate part jointed to the sides. The rest of the can is a single, thin piece of aluminium. Similarly a plastic bottle for carbonated water is cylindrical, except perhaps for the base, with a change in cross section at the top of the bottle. However, in this case the top of the bottle has a screw thread for the cap. I'll return to the shape of screw threads a little later.

The can is not a simple cylinder because its outline (sections along the axis) includes complicated curves. However, the cross sections across the axis are all circles. Things are more complicated in three dimensions (Figure 82). On the body of the can the surface only curves around the can in one direction. On the shoulder the surface also curves around the can but there is another curve up and down. The base is similar to the shoulder.

SAQ 7

Give an example, not referred to above, of cylindrical shapes used in domestic items whose cross sections change.

4.2.2 Spheres

A sphere has the property that any cross section is a circle. Spherical shapes occur as machine parts (for example ball bearings) or as ornamental features such as ball feet on furniture or finials on gateposts. Parts of spheres are more common than whole ones. An example using parts of spheres is the classic design of the Verner Panton lamp (Figure 83).

Figure 83 Parts of spheres on a Verner Panton lamp

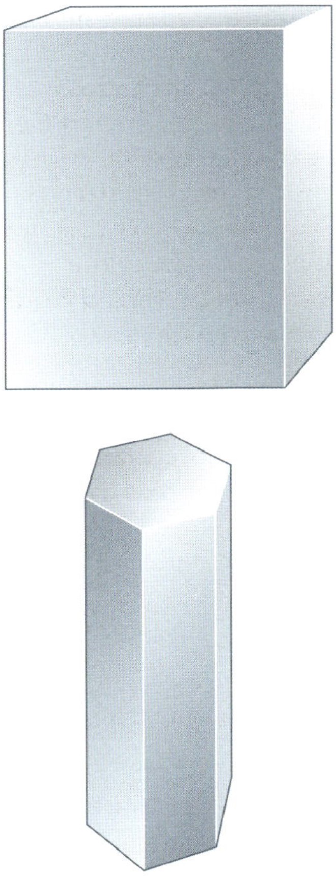

Figure 84 Rectangular and hexagonal prisms

4.2.3 Constant cross sections

Cylindrical shapes have circular cross sections. Similar types of shape have all their cross sections the same shape but not necessarily circles. Prismatic shapes have polygonal cross sections and flat faces. A cereal box is a rectangular prismatic shape (Figure 84), and wheel nuts on a car are hexagonal prisms, though the tops are probably curved.

More complicated cross sections may be wavy or toothed. The cross sections of gears or toothed wheels have complex outlines with peaks and valleys (Figure 85).

Figure 85 Gears

Shapes with even less symmetry but retaining the constant cross section are used for components like table and chair legs or window frames. Moulded UPVC window frame parts are complex shapes that provide strength and weatherproofing (Figure 86).

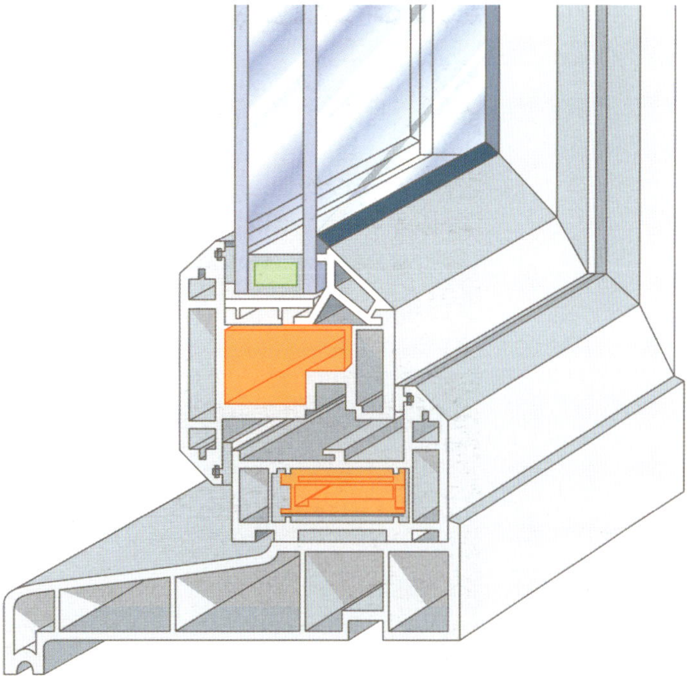

Figure 86 UPVC window frame cross section

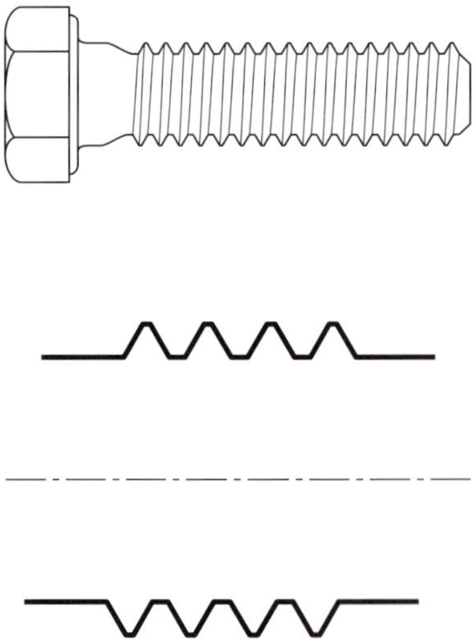

Figure 87 Bolt and its thread profile

Shapes of constant cross section, such as the UPVC window parts, are easy to make. The plastic is squeezed through a shaped hole in a special die tool. A shaped and moulded length of timber in constant cross section, like a picture rail, is also relatively easy to make. With a special set of cutters a rectangular section of wood is fed through the cutters, emerging with a curved cross section.

What do you notice about a shape with a constant cross section such as a cylindrical table leg? The surface curves in one direction and not in the other. However, some constant cross-section surfaces do not work quite like this.

There is one more example of a constant cross-section shape to examine. This is the common screw thread used in nuts and bolts. The screw thread spirals around the surface of the cylindrical bolt and on the inside of the nut.

The head of the bolt has a hexagonal cross section (Figure 87). On the threaded part the cross section along the axis looks like a series of teeth. Across the axis the section is a circle with a projection where the thread crosses the section (not shown).

4.3 Curvy shapes

I suggest that you look again at the DVD sequence on the design of the animal-shaped web cams by Philips. As you view this sequence pay particular attention to the following two things.

First, the animals are configurations of shapes. The body, camera lens (roughly a cylinder) and tail are configured to suggest animal shapes. The sequence on the duck shape illustrates how a reconfiguration of elements created this shape.

Second, the drawings the designer shows of the body of one of the animal web cams have an outline constructed from parts of circles. You can see this on the drawings because different parts of the outline have arrows picking them out with radius values attached to them.

The simple curves I described above are probably outnumbered by curvy shapes. A car is an essay in curves: doors, bodywork, dashboard, even some glass panels, curve in many directions at once. Similarly a bottle of shampoo is often an eye-catching shape, maybe with smooth and complex curves. I'll now look at some examples of these curvy shapes.

You have already seen curvy outlines in the soft drinks can and in the drawing activities in the *Modelling Workbook*. I'll focus on just one aspect of these outlines, namely how curved they are. This property of curvature affects how shapes are perceived as well as how they are manufactured.

4.3.1 Curvature and curved profiles

A circle curves round smoothly until it comes back to meet itself. The circle is curved by the same amount all the way round. A small circle curves more tightly than a large circle. A small circle has a small radius and a large curvature, while the large circle has large radius and small curvature (Figure 88).

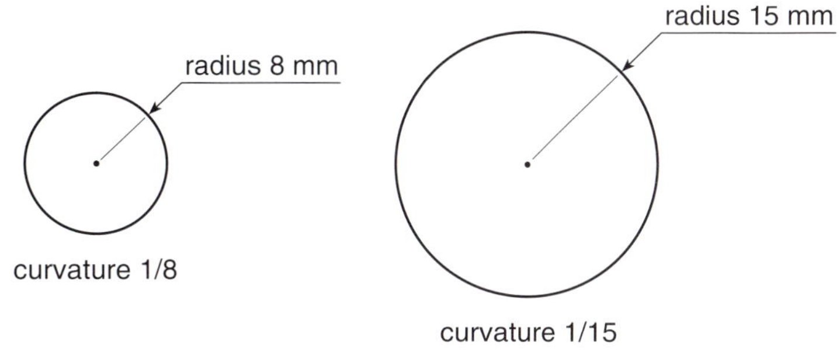

Figure 88 Smaller circle has larger curvature, whereas larger circle has smaller curvature

Now look at a curve made from parts of two circles joined by straight edges. An example is the Ericsson telephone handset illustrated earlier in this block (Figure 23). Figure 89 shows this in a view from the side and reveals its form based on two circles and two lines. Notice the red rim picking out this profile.

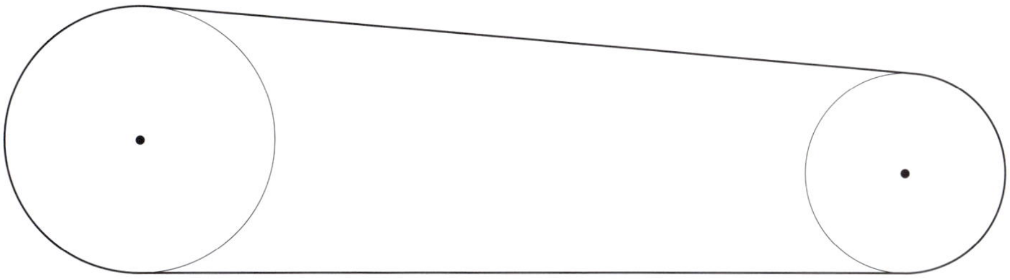

Figure 89 Ericsson phone side view and the circle construction it is based on

Circles and parts of circles are used for many curved shapes. The profile of a tap handle can be constructed as a combination of circles. A section across the top of the tap handle shows a circle in the centre, four circles at the ends of the handle each joined to the centre circle by bits of two large circles (Figure 90). A simpler example is the link on a bicycle chain. This has two more circles for the holes for rivets to join it to the next link.

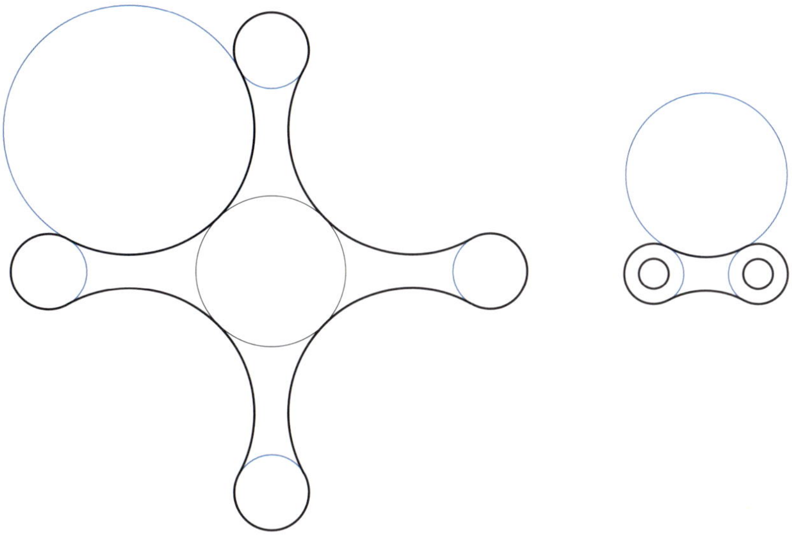

Figure 90 Tap handle and bicycle chain link with the circles they are constructed from shown

The bits of circle used to construct the profile of the tap are joined smoothly so that there are no sharp corners.

In the profile of the tap handle the curving line goes in and then out again. When the circle encloses part of an object the curvature is positive and the shape is called convex. The centre and the four handle ends of the tap shape are convex. When the circle eats into the solid then the curvature is negative and the shape is called concave. The arms of the tap handle have a concave profile.

Circles are widely used as parts of profiles. Profiles may be combinations of circular arcs joined smoothly together but many products have more complex outlines and shapes.

Exercise 18 Concave or convex

The profiles shown in Figure 91 are picked out by circular outlines. Which parts of the profile are concave and which convex? How are the parts of the circles joined together to form the whole profile?

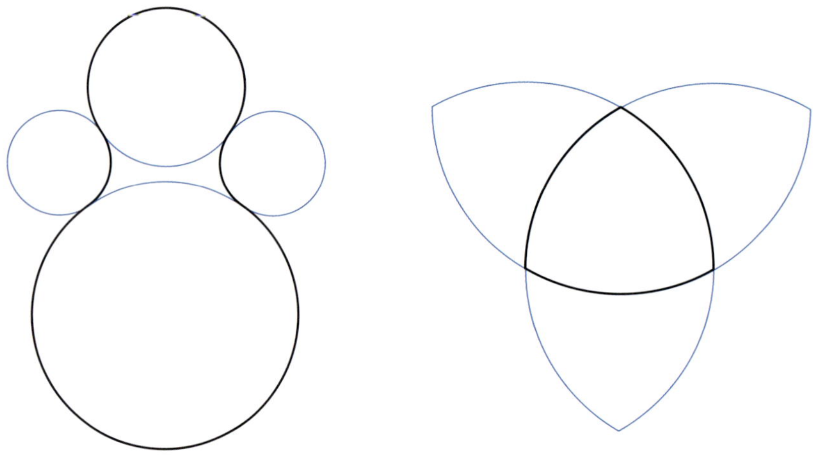

Figure 91 Smooth guitar-like shape and curved triangle with the curves they are constructed from

Discussion

In the first shape the top and bottom are convex and the two dips at the sides are concave. The ends have smaller curvatures than the other parts. The parts of circles are joined smoothly. There are no corners at the joins.

In the curved triangle there are sharp corners at the joins and the curves are convex.

4.3.2 Defining complex curvy shapes

What do you think it means to say that a curved shape is *not* circular? You have seen that a circle curves round smoothly. In fact a circle always curves around by the same amount until it comes back to meet itself. Curves that are not circular change their curvature as you trace along their length.

The tap handle section and the Ericsson phone each have three different curvatures. In the phone the straight-line parts of the outline have zero curvature and the two ends have different curvatures because the circles are different sizes.

How would you define the curvature of an object? Informally and imprecisely, if you traced along a curve and found it changed direction a lot in a short distance, you could say it had a high curvature. This can be expressed mathematically but it is enough to notice that a large radius circle turns slowly as you trace it and a small radius turns quickly. Curvature is the reciprocal of the radius of the circle – expressed as 1/radius. A circle with radius 20 mm has a curvature of 1/20 = 0.05 (see Figure 88).

Although a few curves without constant curvature can be drawn by hand quite easily they are usually difficult to draw accurately. This is why, before the availability of CAD, designers used standard curves such as those you saw earlier in Figure 77.

Figure 92 Curve with smooth changes in curvature (left), and curve with an abrupt change in curvature (right)

The Ericsson phone displays a smooth, even transition from the flat to the curved areas. However, a join between a straight line and a circle of radius 10 mm is not especially smooth because the curvature changes from zero on the straight line to 1/10 = 0.1 on the circular arc. Similarly if two circles are joined with different radii as in the tap, curvature will change abruptly at the join. One way to achieve a particularly smooth curve along a long length is to ensure there are no abrupt changes in curvature (see Figure 92). CAD software can easily create these smooth curves because it uses mathematical descriptions of the curves.

Although the profile of a product is an important feature, it is really the surfaces of the product that define its shape. Surfaces that contain curves with changing curvature – which are not made up from circular arcs – can be difficult to manufacture. These curves match human shapes and movement, give longer life because shapes can be stronger, have attractive appearance and can create product identity. These enhanced functions can justify the costs and difficulty of making these shapes.

These curvy shapes can be seen in chair designs. The Verner Panton chair in Figure 93 is a single shape with many different curves. It is moulded in plastic. You will notice this shape is symmetrical about the centre of the chair. Looking at its profile from the side, you see a shape that starts narrow at the top, becoming wider at the seat and progressively narrower at the front leg. The foot space and rear support makes a bracket shape at the base of the chair. The front and back of this profile are distinct curves. The profile gets wider at the bottom of the back and on the seat because these carry a weight and are likely to be heavily loaded as a person leans back in the chair. Sections across the back and seat of the chair are also made up from complicated curves. These curves provide a shape to support a person. They also give strength, a dynamic appearance, and a product identity

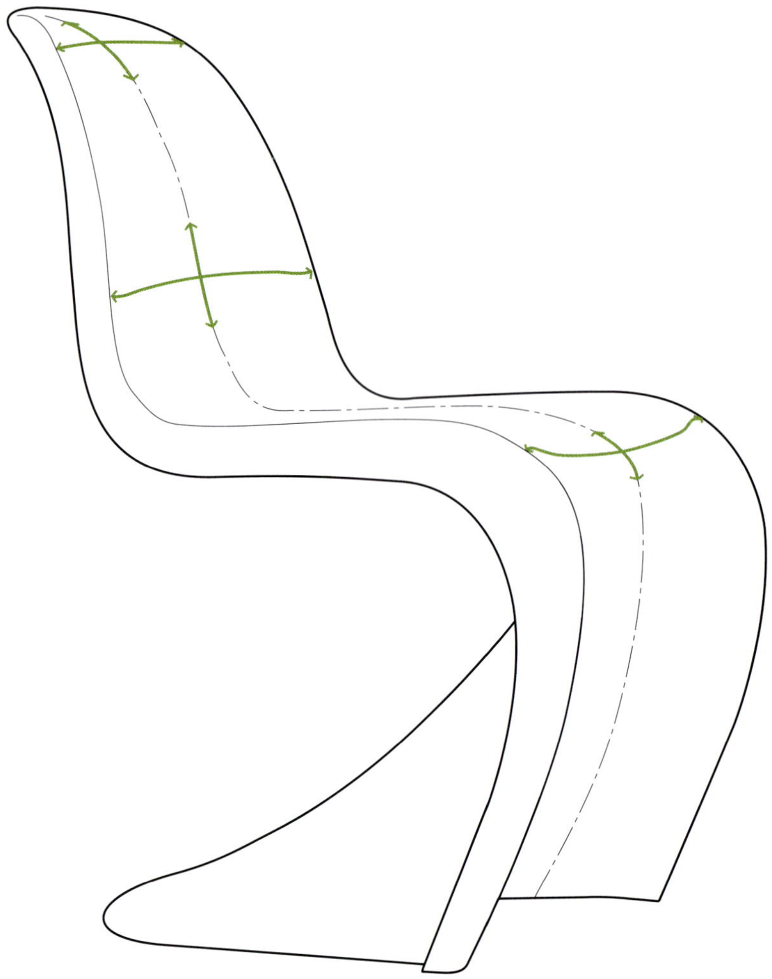

Figure 93 Verner Panton chair is a good example of curvy surface

How do designers describe these complicated surfaces? Two main methods are used. In one method they build accurate models, altering and varying them until precisely the right shape is created. Traditionally car designers built highly accurate, full-size clay models by careful shaping and scraping. When the exterior form was acceptable, measurements could be taken from the surfaces. More recently such clay models facilitated the digitisation of geometry for subsequent working on computer.

More recently still, design teams construct the surfaces and forms they want directly in CAD programs, such as the one supplied on the course DVD. However, these CAD models have not replaced the full-size accurate and physical clay models of cars. Both computer and physical models are used side by side. Complicated surfaces can be created and manipulated on the screen. Surfaces can be stretched and shrunk, twisted and flattened out. Different lighting conditions can be simulated to show surface finishes such as glossy, patterned or textured.

Most shapes, with their visualised surfaces, don't stay in the computer. Clients and designers need to see and feel the shapes and surfaces so prototypes are produced from the digital data of the computer models. These prototypes are made in several ways, depending on the type of product. In the fashion industry, actual garments will be made. For consumer products, rapid prototypes will be made that mimic many of the characteristics of appearance and feel of the final product.

4.3.3 Properties associated with curves

How do curves give strength? Look at Figure 93. At any point on its surface the Verner Panton chair curves in all directions, usually by different amounts. On the top surface of the chair at the edges of the seat there is an S shape from the top of the back to the front of the seat. If you look at a section across from one side of the chair to the other, you will notice a roughly M-shaped curve. The combination of these two curves one from top to bottom and the other from side to side give the chair its strength – and contribute to its distinctive appearance.

The Verner Panton chair is an extreme example of a curvy shape. The curves on the surface of the chair turn this way and that. This helps to prevent the shape twisting as someone adjusts their position in the chair.

The general principle is that curves in the surface help to strengthen a sheet. The panels on a car are curved not just for an attractive appearance. Ribs or corrugations stop the panels flexing. Just as in the chair, car panels have gentle curves as well as tight curves. Roughly the gentle curves give a smooth looking shape while the tight curves give strength. In a car the tight curves may be hidden or half-hidden where the doors and bonnet close. They are often near the edges of panels. But the strengthening curves should not be too tightly curved because they then become hard to manufacture. Tight curves are also places where the material can start to crack.

SAQ 8

Have another look at the soft drinks can that you examined earlier in this section (Figure 82). Every part of the surface is curved. Which parts curve in more than one direction? Identify the parts of the surface with the highest curvature. Try to answer these questions before looking at Figure 94.

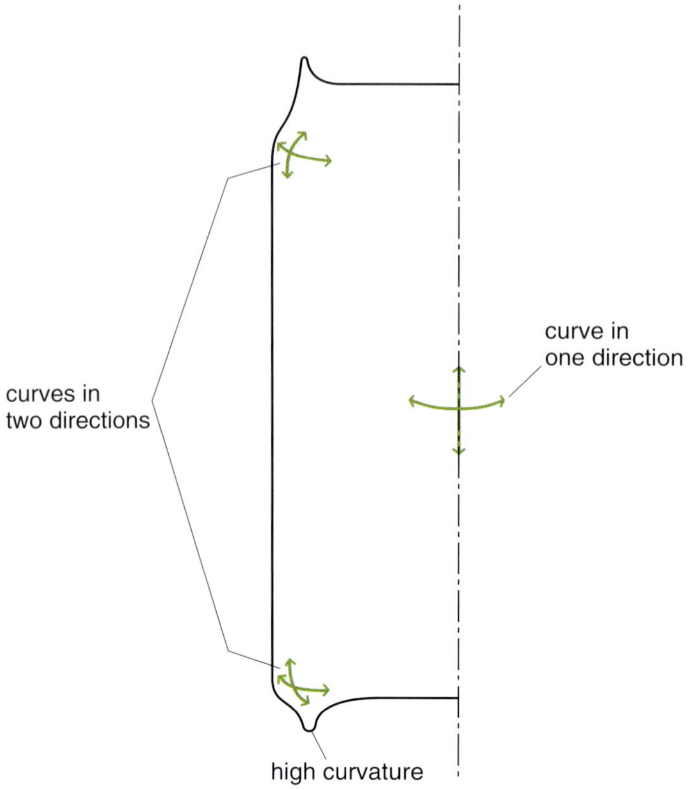

Figure 94 Curvatures on sections of a soft drinks can

The example of the moulded Vernon Panton chair shows clearly that shape contributes importantly to usability, aesthetics and product identity. Shape also contributes to reduced environmental impacts. For example, the aerodynamic profile of a vehicle saves energy, or the thinner steel with strengthening curves reduces vehicle weight.

To deliver functions, shapes must be specified exactly. A small inaccuracy in a dimension might mean that gears do not mesh properly and wear quickly. A tiny error in describing the curves in a car body might mean that parts do not fit smoothly together perhaps ruining the appearance. And it is vital to specify dimensions precisely for manufacture and quality control.

W 2

This is a good point to read Section 6 of the *Shapes and Space Workbook*, 'Where do curves come from?'

4.4 Component dimensions

To describe a shape precisely you need to specify its dimensions. What dimensions are required to describe the various types of shape you have already seen in this section?

4.4.1 Cylinder dimensions

I've already described how to specify a cylinder. The dimensions of its length and radius are needed. The dimensions of the cylinder shown in Figure 95 are a radius of 10 mm and a length of 80 mm. Notice the way the cylinder is drawn as two views, one from the top and the other from the end.

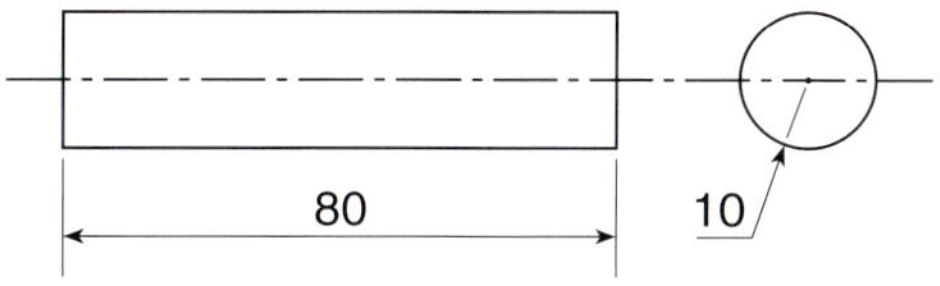

Figure 95 Dimensions of cylinder

For general cylindrical shapes whose sections change along their length, like the shaft of an electric motor, a few more dimensions are needed. Recall the shaft of the washing machine motor in Figure 80. It is necessary to give the radius of each cylindrical part and its length. These dimensions are shown on the drawing of the shaft. A dashed line is shown along the axis of the shaft. Arrows pointing to each cylindrical part indicate the radius of that part (Figure 96).

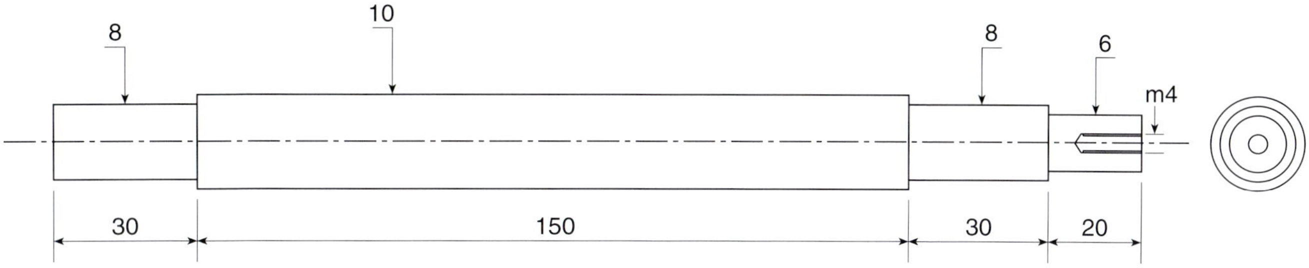

Figure 96 Dimensions on the side view of an electric motor shaft

4.4.2 Corner and profile dimensions

For other cylindrical shapes whose radius varies continuously it is necessary to describe the profile. Figure 97 shows the enlarged end of the shaft that was in Figure 96. On the steps you notice a rounded concave corner. Apart from the fact that manufacturing a sharp concave corner is difficult, it would not be sensible. The sharp corner would be liable to be the starting point of a crack. The rounded corner goes all around the cylindrical shape. As you saw before, the simplest shape for the rounded corner is a part of a circle. To specify this corner you just have to say what the radius is to be.

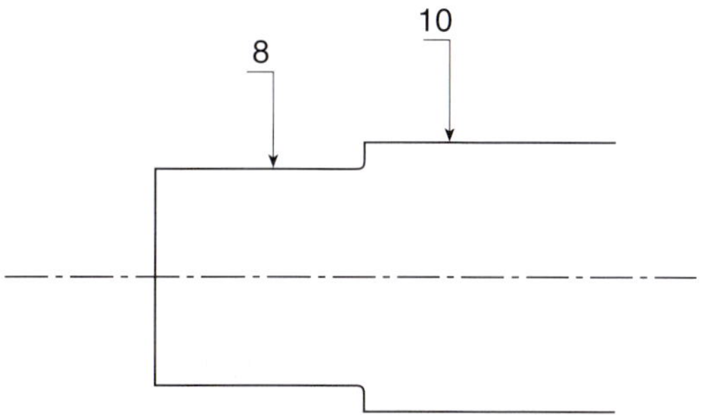

Figure 97 Rounded corners on cylindrical component

Spherical shapes are easily specified by a single radius. If part of the sphere is cut off, the position of the cut needs to be specified. Figure 98 is a sphere with a part cut off and the diagram is how the cut would be specified.

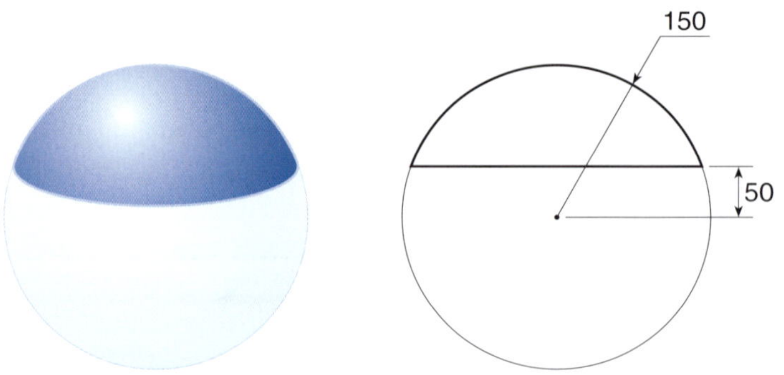

Figure 98 Cut sphere

4.4.3 Specifying dimensions

More complex shapes like gears or toothed wheels are not usually drawn in detail with all their dimensions. The reason for this is that shapes like gears have standard shapes that are defined for all gears. National and international standards organisations such as the BSI, Deutsches Institut für Normung, or International Organization for Standardization will specify exactly what the geometrical shape of a gear tooth is. The standard refers to particular key dimensions for which the whole shape of the tooth can be derived. These key dimensions will include the radius of the gear (radius of the pitch circle), the number of teeth and their size (module).

Other examples of circular shapes with repeated features are the wheels on a car. At the centre of the wheel you see a central hole with four or five wheel stud holes spaced evenly (Figure 99). How is the shape with five stud holes specified?

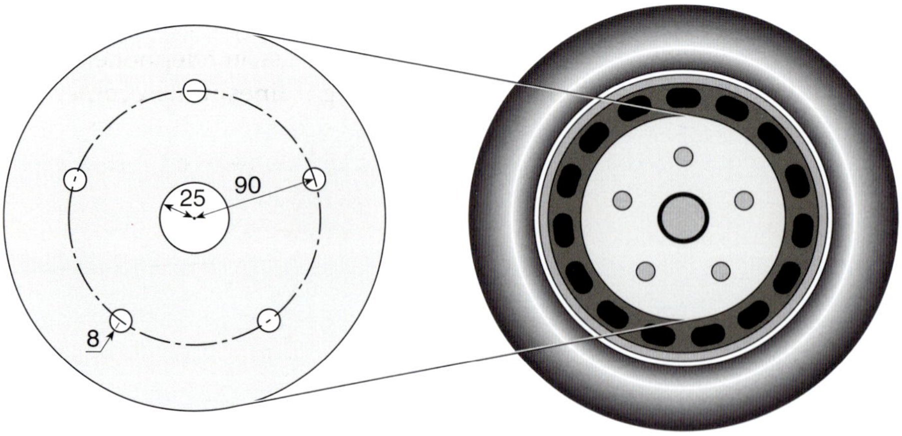

Figure 99 Car wheel with stud hole positions

First, you could specify the radius of the hole in the centre. Then you could specify the radius of each of the stud holes – these are all the same so it's only necessary to specify one of them. Defining the position of the stud hole is a little more tricky. The centres of the stud holes lie at a certain radius from the centre of the wheel. This is specified. Finally the stud holes are evenly spaced so there is an angle of 72 degrees (360 divided by 5) between them.

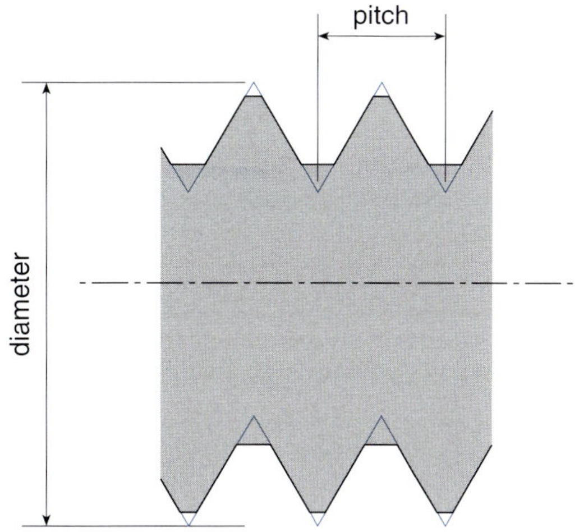

Figure 100 Screw thread diameter and pitch

How would you specify a spiral or screw thread? The thread goes round and round the cylindrical surface of the bolt (see Figure 100). As it goes round it also moves along the bolt. For each turn right round the surface it moves along by exactly the same amount. This distance is called the pitch of the screw thread. So to specify a screw thread you need two numbers, the diameter of the cylinder and the pitch. The shape of the thread also needs specifying but this is done by referring to standards, like the gear teeth mentioned before.

4.4.4 Curvy outlines and surfaces

In computer-aided design models, curves and surfaces are defined by joining small pieces of simpler curves or surfaces. You saw straightforward examples of this that used parts of circles. To specify these it is necessary to give the position of the circles by the location of their centres, their radii and the extent of the circular arcs. The profile of the Ericsson telephone shown earlier has two circular arcs and two straight lines in its profile. The whole shape is specified by the sizes of, and the distance between, the two circles. All the other dimensions can be derived (see Figure 101).

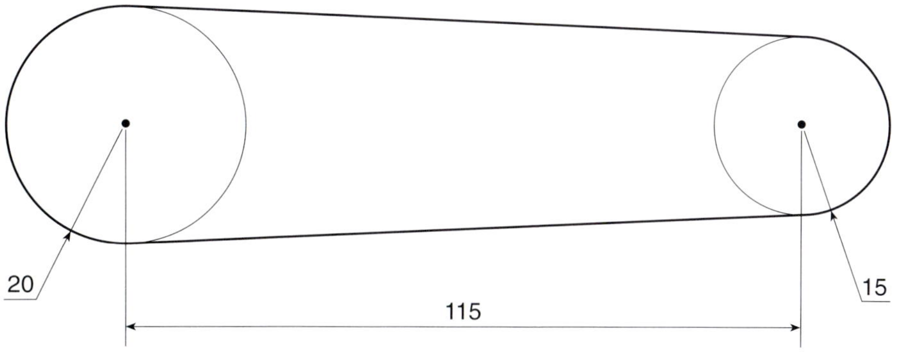

Figure 101 Specifying the profile of Ericsson phone shown in Figure 89

Figure 102 iPod digital music player

SAQ 9

Look at the picture of the iPod in Figure 102. How would you specify the outline of this front view?

Designers use both physical models and CAD models to define the shape of components. When a designer first creates physical models these may be transferred to computer models by digitising the surface. When computer models are created first, physical models may subsequently be generated by rapid prototyping techniques.

To get the shapes of physical models into the computer takes time. Essentially measurements of the positions of many points in 3D space need to be determined. In car styling, thousands of points need to be defined for each area of the surface of the clay model. These then become the coordinates that allow the shape to be accurately reconstructed in digital form. Special CAD programs are used to fit surfaces to all the points. These accurate models of the shape can then be used to make drawings, to analyse the strength of the component, and to make the moulds that will produce the actual component.

4.5 Strength in components

Strength is one of the most important properties of the components that make up the consumer products that surround us. Without sufficient strength products can fail. You expect chairs to be strong enough to bear your weight, knives and forks not to bend and the components of a car hoist to be strong enough to lift people in and out of cars again and again without breaking or bending.

Strength seems a straightforward idea but as with many things in design it turns out to have many facets. What is meant when someone says a component is strong? A seat belt in a car is strong when you pull on it but it also rolls up easily on a reel. The seat belt resists tensile loads. A beam lintel over a door or window in a house only bends very slightly under the load of stones or bricks above, so the lintel is strong because it resists bending. The lintel is not quite so strong in resisting a twisting load but that kind of load would be abnormal. A soft drinks can holds the carbonated drink under pressure and is firm to hold when unopened. However, it can be crumpled easily when it is empty. A cardboard cereal packet bends and distorts easily, but it does not need to be strong as its purpose is temporary and mainly concerned with appearance.

4.5.1 Material and shape affect strength

Components are strong in ways appropriate for their use and their strength is a consequence of many factors. Clearly the material from which a component is made is a major factor – steel forks are stronger than plastic forks – but shape also determines properties of strength. Many of these properties are about how a component responds to different types of load. I noted above that a lintel might strongly resist bending but is less strong when given a twist. A car seat belt resists a pulling load but bends easily to wind up on a reel. Be aware there are many different types of load – pulling, compressing, bending and twisting for example.

So the responses of a component to any load depend largely on its shape and material. A woven seat belt is quite wide to spread any load and is joined to a steel tongue that goes into the buckle. In the buckle a rectangular steel pin clicks up through the hole in the steel tongue to hold the belt in the buckle (see Figure 103).

In a collision the pin and the tongue pull on one another. The pin must not break off and the material around the hole in the tongue must not crack. With a reasonably large contact area between the pin and the tongue, a strong pin, enough material around the hole and rounded corners, failures like cracking and breaking can be avoided. In fact the designer must make sure they don't occur at all. Just one failure in many millions of seat belts is one too many.

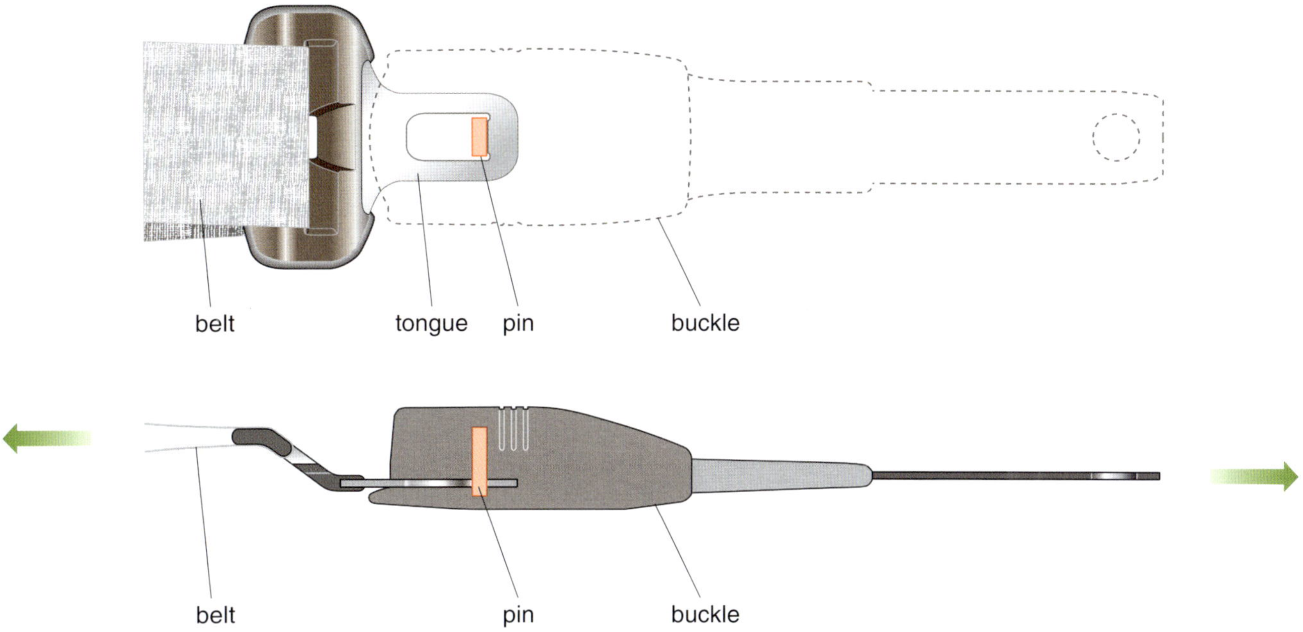

Figure 103 Seat belt components and loads

What principles can you learn from the seat belt tongue for making a strong component?

1. Put plenty of material around where the forces are greatest.
2. Where there is contact, and one component pushes or pulls on another, try to spread the contact across a large area.
3. Where there are large forces make sure any corners in components are rounded.
4. Use a material that is strong.

In the seat belt there is a limit on the amounts of material and area of contact. To make up for this a strong steel is chosen. But what does a strong steel mean? It means it resists loads without breaking or being deformed permanently. These loads include pulling, compressing, bending or twisting.

Special testing machines are used to find out what load is needed to break or permanently deform materials. Specially shaped samples of material are subjected to measured loads. A common strength test is one where a pulling load is increased until the sample breaks.

The pulling load to break a sample will depend on the size (area of the cross section) of the sample. It turns out that twice the area needs double the pulling force to break it. This confirms the first and second principles above.

Now I will look at the lintel to see if there are any more principles. Here the load of the wall above tries to bend the lintel. It should only deform slightly. Although the lintel is quite large it does not have a lot of material in it – it is hollow (Figure 104). These lintels work well and illustrate another principle.

When a designer wants a component to be strong against bending they use a shape that puts as much material as possible near the outside of the component. Hollow components – including shell-like shapes – can be good at resisting loads.

Figure 104 Hollow steel lintel

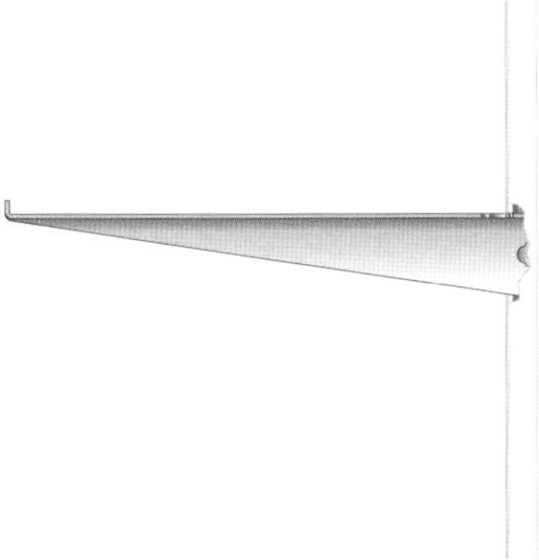

Figure 105 Bookshelf bracket needs more material near fixed end and less at the tip

It is worth mentioning one further design principle that helps to make a component strong. A component fixed at one end with a load along its length, like the bracket on a bookshelf, can taper towards its end (Figure 105). The fixed end should have more material than the tip. The taper will be determined partly by the material from which the component is made, the length of the component and the load it has to bear.

So there are two more principles to add to the list above.

5 Hollow components can be light and strong. To achieve strength, material is placed near the outside of a component that is subject to bending, like the tube in a bicycle frame or the steel lintel.

6 Tapering, cantilevered components can reduce material content while still providing strength.

Exercise 19 Hollow but strong

How much compression load can a sheet of A4 paper withstand when it is rolled into a tube? You may well say it depends on the material and how many layers of paper make up the tube, and you would be right.

Take a sheet of A4 paper that is typical of photocopiers or computer printers – it is usually called 80 g/m^2 paper. Roll this up to make a tube that is 210 mm long and about 50 mm diameter. Stick some tape on the outside to stop the paper unrolling.

Stand the tube upright on a smooth, level surface. Move anything breakable well clear of it. Put a paperback book horizontally on top of it so the book is balanced. Keep adding books or video tapes or similar objects on top of the first and try to keep them balanced. When the tube failed under the compressive load what was the load?

Discussion

My tube withstood 11 paperback books and failed when I put the twelfth on. I think quality of manufacture played a part. I only put a short piece of sticky tape on the outside in the middle of the seam. I think taping the whole length of the seam on the outside of the tube would have made it stronger. These kind of empirical tests help designers understand the properties of components.

This was a failure due to a compressive load along the length of the tube that eventually caused the tube to fail. With tubular bicycle frames, for instance, there may also be bending and tensile loads on the tubes.

The three examples of the seat belt buckle, the lintel and bookshelf bracket indicate how both the shape and material properties of a component combine to provide strength.

In both the seat belt buckle and the lintel the component should be rigid. The seat belt fabric should be flexible for winding on to the reel and should be strong against a pulling load.

4.5.2 Components can be elastic, resilient, tough, hard or dense

Perhaps without realising it you rely on some components that are somewhere between rigid and flexible. A spring changes shape – it is flexible, within limits. The spring stretches or compresses by an amount that depends on the pulling or pushing load. Double the pulling load makes the spring extend twice as much. All components act like springs to some extent and deform under loads. If the load is not too great, a component returns to its original shape when the load is released. The deformation is said to be *elastic* – the component behaves like a spring. If the component that is subject to a load does not return to its original shape but is permanently deformed, that is called a plastic deformation. The way a component deforms depends on both its shape and the material or materials it is made from.

One material will vary in its 'springiness' from another material. The springiness is more accurately described as elasticity. You can compare materials by referring to published tables, which indicate elasticity by showing how much load has to be applied to a contact area to cause a given deformation.

Special terminology is used to describe elastic deformation. A material that deforms easily under a small load is like a soft spring and is said to have a low *modulus* (a low value) of elasticity. For example, rubber deforms under a small load and springs back into shape. On the other hand a material that is like a hard spring, deforming only slightly under a large load, is said to have a high modulus (a high value) of elasticity. Examples include glass, aluminium and steel. Tables are published that show values for modulus. Do you think the modulus for diamond is closer to rubber or closer to steel?

Another material property can be defined that combines the properties of elasticity and strength. A component with relatively high strength in relation to elasticity is called *resilient*. Roughly this means a component that is resilient can absorb loads by elastic deformation.

To illustrate strength and elasticity I'll return to the two forks. If one is made from stainless steel, it is quite heavy. The fork deflects slightly as I use it but it returns to its original shape. This is a relatively strong material with a quite high modulus of elasticity. However, I could probably bend the fork permanently if I try hard enough, especially at the thinnest part of the handle.

If the fork is made from plastic as a low-cost disposable item, it is much lighter in weight. It will not resist as much loading as the steel fork – it is not so strong. But bear in mind it is lighter, so weight for weight it may be stronger than it seems at first. The plastic fork deflects quite a lot under load and has a smaller elasticity than the steel. In addition the plastic fork breaks while the steel fork bends under larger loads. Therefore the plastic fork is less resilient than the steel fork.

If the deformation is elastic and the component returns to its original shape when the load is released, it is likely the component can still function as intended. Indeed in many products this is a key property. A washing up detergent bottle is deformed by squeezing every time it is used but then returns to its original shape. Alternatively some permanent deformation can be desirable. A tube of toothpaste is a good example. However, for precisely engineered components, such as those in a gearbox, designers are careful to avoid permanent deformation – if gear teeth become misshapen, they cannot mesh correctly. In many products small amounts of permanent deformation can be tolerated provided components do not break.

Failure by breaking is complicated. Sometimes breaking occurs quickly – the component cracks and breaks apart, in other words it fractures. You might be most familiar with glass or ceramic materials, which are brittle and tend to crack easily. The disposable plastic fork is not as brittle as glass. It can bend but then it suddenly cracks and breaks quickly.

If a material does not crack easily, or more accurately if the cracks do not grow, it is said to be *tough*. The disposable plastic fork is less tough than the steel fork, which resists loads, bends a bit in response to loads, but is unlikely to break. The seat belt buckle has a pin that is strong and tough because it must resist large loads and definitely must not crack and break.

Tough components may be strong but not all are strong. You have probably tried to open with your fingers the clear plastic film used on food packaging – it tends to stretch and deform without tearing. Even after introducing a crack by cutting with a knife, some plastic films still resist tearing, that is, the crack will not grow. This is a tough material that is not particularly strong – it deforms easily but will not break.

If a component deforms under a load, the deformation can spread loads around a component and prevent the creation and growth of cracks. So toughness tends to be associated with a bit of 'give' in all directions. Ceramic cups and plates in the kitchen do not give, they are brittle and break suddenly especially when they suffer a sharp knock.

A further property is closely related to strength and toughness. Some materials are *hard* and their surfaces do not scratch easily. Hardness is a measure of response to a particular kind of load on the component. This is the load from a sharp object pressed into the surface. Press lightly and the surface will spring back but press more forcefully and you make a mark in the surface. Strong materials tend to be hard as the surface does not deform permanently – it does not scratch. Glass is hard and does not scratch easily. Plastics are generally less hard and scratch more easily.

Finally I mention another material property that can be important in design, namely *density*. This is a material's weight for a given volume. In the example of the steel and the plastic table forks you saw how the plastic fork was considerably lighter for the same size, but unfortunately did not perform as well especially in regard to strength and toughness. In many designs lightness is a great advantage, certainly in cars, aeroplanes and portable products. Density is not important by itself – in an aeroplane frame, components should be light but they also need to be strong and tough.

4.6 Surface finish and texture

Having designed a component to be the right shape, size and strength you may want to make it look and feel attractive. This will involve you in a consideration of the properties of surfaces, namely their finishes and textures.

Components in a product can have surfaces that are the result of one process or several processes. Some finishes on a surface will be the direct result of a manufacturing process and the material from which the component is made. For example, the finishes on the plastic casing of a vacuum cleaner, a car dashboard or the fabric in a garment come straight from manufacture. The quality of their surfaces will be determined by the machines and processes used to manufacture them. Being able to use components straight from the manufacturer is an advantage because the cost and time in preparing the surface is avoided. Other products display surfaces that require many stages of preparation.

Surface finishes can be textures in the surface itself or applied to the surface as coatings. These finishes give a special appearance as well as a feel to the component and they can contribute to other functions. Surfaces in bearings that must slide or roll over each other are polished to reduce friction and hardened on the surface to resist wear. Protective coatings, like paints prevent corrosion and prolong life.

4.6.1 Smoothness

I will focus on the appearance and feel of a surface. A surface that is polished looks bright and feels smooth. By polishing, small irregularities like humps and dips are levelled off. The roughness of a surface can actually be measured in terms of the sizes and shapes of these humps and dips. A smooth surface has a very small distance between the tops of the humps and the bottom of the dips (Figure 106).

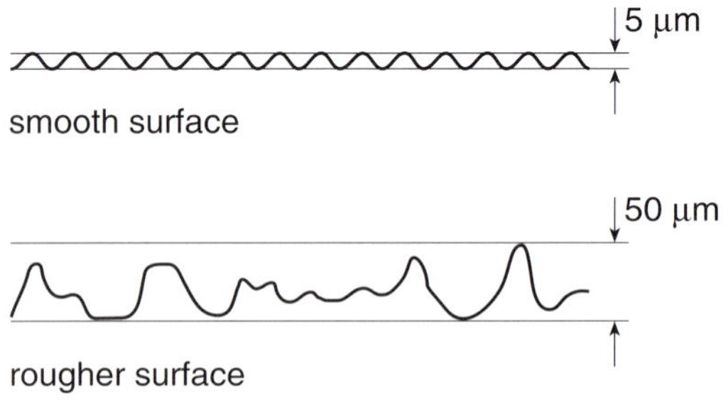

Figure 106 Smoothness of surfaces is measurable (the unit µm is a micrometre – one millionth of a metre)

If the humps and dips are small but jagged, there is another kind of roughness. This might be measured by the tightness of the curves on the humps. Small radius, jagged humps produce a rough surface such as that you can feel on a house brick.

How can you polish a surface to make it smooth? It can be polished mechanically with a wheel or the backwards and forwards motion of a pad. Usually a mild abrasive is used to help remove the humps on the surface. Polishing is done in stages to get an increasingly smooth surface, starting with a course abrasive and ending with a fine abrasive. Mechanical polishing takes a long time and costs a lot so is avoided except where absolutely needed for high precision surfaces such as lenses and mirrors for optical instruments.

Another kind of polishing removes the top layer of the surface by dissolving it chemically. This relies on the humps and peaks being acted on by the chemicals faster than the dips and troughs. This is a common method of polishing to obtain a smooth appearance for the visible parts of products that are used everyday but that don't have to be precise.

4.6.2 Coatings and surface treatments

A major type of surface treatment is painting. These coatings may be protective or decorative or both. The exterior surfaces of drink cans are printed decoratively with colours and even varnishes for a glossy appearance, while their interiors are coated protectively to prevent taste contamination from metal to drink. There are many painting, printing and surface decoration processes and I will not deal with these except to remark that a huge amount of design activity is targeted at them. The manufacturing processes are quite amazing and designers are always exploiting the possibilities offered by these processes. Aluminium drinks cans are produced in large numbers, and each can needs to be printed with graphic designs. A single printing machine prints seven colour designs onto 2000 cans per minute.

But surfaces are not always polished and glossy. Surfaces can be specially treated by processes such as anodising to give them a matt finish. The technical details are not important here. This is often applied to aluminium and consists of promoting the growth of an oxide layer on the surface by immersing the component in acid through which an electric current is passed. This protective coating can absorb colouring pigments easily, so coloured, anodised finishes are quite common. You may have seen some examples of modern jewellery that exploits this anodising process, and I show some engineering components in Figure 107.

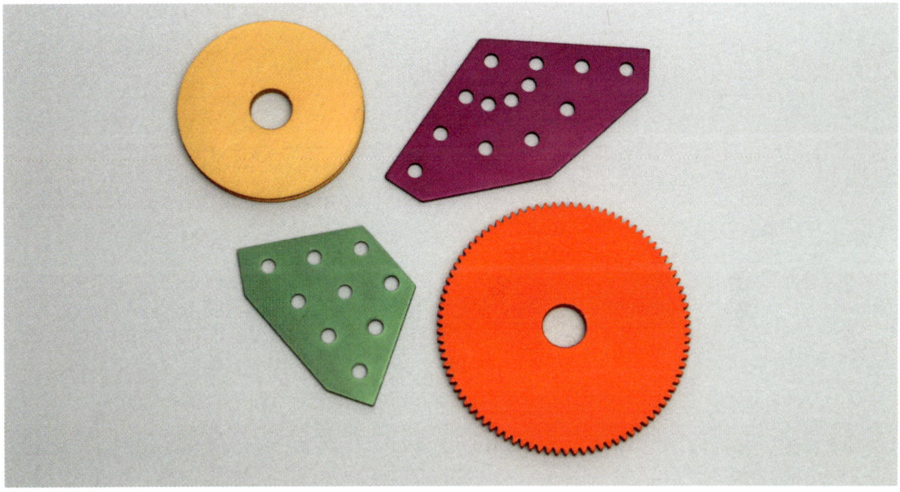

Figure 107 Anodised surface finishes

If you've ever done any home decoration, you will probably have come across paints with finishes such as matt, silk, eggshell, and so on. But what is a matt surface? It can feel pleasantly soft if you run your hand over it and does not reflect bright spots on its surface. Light from lamps or sunlight is diffused by the surface, which appears the same when viewed from different directions. This is not the case with a glossy surface, which reflects a bright light. Different surfaces will reflect bright lights in different ways. These are important visual features that can have important safety implications, for example in the visibility of instrument panels.

4.6.3 Virtual surfaces

Computer-aided design systems can visualise realistic pictures of surfaces under different illumination. The positioning of lights, the colour of the light and the reflective properties of surfaces can be varied to evaluate surface appearance. In this way designers can obtain a realistic visualisation of their products as they work on the details of surface finish and colour.

But these CAD visualisations do not allow designers to feel the physical presence of a component. Now, virtual reality technology allows you to touch and interact with computer models, usually by placing your hand in a special glove that is located in space by sensors. The glove responds to finger movements and exerts forces on the fingers to give the impression that the user is touching a physical object. In this way these virtual objects can be moved or picked up. A related technology was shown by Mark Evans in his essay in the *Modelling Workbook*. In this case a pen-like device allowed designers to feel the surfaces of their virtual product.

Exercise 20 Feel the difference

Run your hand across several familiar objects such as a tabletop, the screen of your computer monitor, the surrounding casing of the screen, a garment you are wearing, the cover of this book and the surface of a ceramic plate.

Try to describe with a couple of words the feel of each item. You can feel alternative surfaces to the ones I've given. Feel anything around you at the time you are reading this – though you may get some odd looks on the bus or train.

Discussion

Could you describe differences in surfaces? You may have noticed metals and glass conduct heat away from your hand and therefore feel cold, while plastics and materials that don't conduct heat feel warmer. A tabletop of wood would have given a grainy texture – smoother in one direction that the other. A glass computer screen would have felt smooth but the surrounding casing would probably have felt slightly uneven. Ceramic plates would have been very smooth and rather cold.

I tried this exercise when I was sitting in my car because its interior has lots of different surfaces. One thing I noticed about the inside of the car was the number of surfaces that are quite soft. Some had 'give', like the seats, while others like the dashboard were harder. Some items possessed combinations of hard and soft parts like the steering wheel. The dashboard had a distinct texture, which seemed to have been done partly to avoid light glare.

4.7 Colour

Colour can have many purposes, both practical and psychological. Some of the psychological uses of colour are discussed in the *Shapes and Space Workbook*. Here I'll look at some of the purposes it can serve when applied to components.

First, it can be used to direct attention at particular parts of a product. In some cases the contrast is subtle while in other cases parts contrast strongly with other parts. Take, for example, the intense colour of the name printed on a packet of detergent or even a distinctive background colour like the design of a chocolate bar. Alternatively this capability is useful where one component needs to stand out from its surroundings, for example the bright colouring of safety features. A red off button on a machine is easily seen so that it can be used quickly.

Second, different colours are used to distinguish different parts of the product. An example you have seen already is the Dyson vacuum cleaner, which incorporates brightly coloured components partly as product identity and partly to communicate, for example where to press, where to open or where to hold.

Third, similar colours or the same colour is used to link separate parts or to emphasise associations. Areas of colour on a product are more identifiable than the lines separating the parts. Using this technique, designers can link up spatially separated parts of a product, perhaps to indicate a similar function or to unify a complex shape. Body-coloured bumpers were introduced on cars to unify a complex assembly of parts and offered new opportunities for body styling in cars.

Fourth, colour is used in practical ways to code electrical components, wires or utility pipes. Gas, water and fibre optic cables are laid underground with yellow, blue and green coverings. Similarly colour is used in toys to make strong distinctions between parts. Figure 108 shows four wires in a cable distinguished by colour and blocks used by young children in four distinct colours.

Figure 108 Colour used to make strong distinctions for serious reasons (electric safety) and for leisure reasons (stimulating toys)

4.7.1 Colour perception and definition

How does the eye see colours? The colours are a response to different kinds of light. Light can be thought of as electromagnetic waves moving through space and the different colours of light are created in the brain as it responds to different wavelengths reaching the eye's retina. A mixture of responses from all the wavelengths in the light are combined to give a single colour to the light.

The eye is only sensitive to a range of wavelengths. You can think of light from the surface of a component as a mixture of different amounts of light of different wavelengths. The way the brain perceives the colour of a component depends on this mixture. Designers, artists and scientists have used a variety of ways to describe the colours of objects. One of these ways describes a colour has having three characteristics:

- the colour type or *hue* ranging through red, yellow, green, blue (Figure 109);
- the *lightness* ranging from light to dark;
- the strength of the colour, ranging from subdued (dull) to saturated (vivid, intense), which is called *saturation*.

Figure 109 Range of hues

The strip of colours in Figure 109 shows a range of hues starting from a red on the left and ending with a similar red. This range can be put into a circle, with the similar colours at the beginning and end put next to each other. Notice the range of hues is different from the spectrum of colours in a rainbow. A spectrum shows light split into wavelengths, while the range of hues shows mixtures of wavelengths.

The terms hue, lightness and saturation are terms used by colour scientists, artists and designers to describe the experience of colour. I think you will be able to distinguish the broad differences between hue, lightness and saturation but it may take a little time to be able to apply the terms in a meaningful way.

To help you, I can place more familiar terms alongside these technical terms. For example, to say a colour is paler may mean it is lighter and less saturated; to say a colour is deeper may mean it is darker and more saturated; to say a colour is more brilliant may mean it is lighter and more saturated. Describing a colour can be confusing, which is why designers refer to the three basic terms of hue, lightness and saturation and specify colours by using a value for of each of these.

Colour charts setting out selections of possible colours are organised according to three values. Some charts use hue, lightness and saturation but others use different sets of three characteristics. The Munsell or Pantone colour charts are commercial systems used by designers in selecting and matching colours. A general chart is shown in Figure 110 using the three values of hue, lightness and saturation.

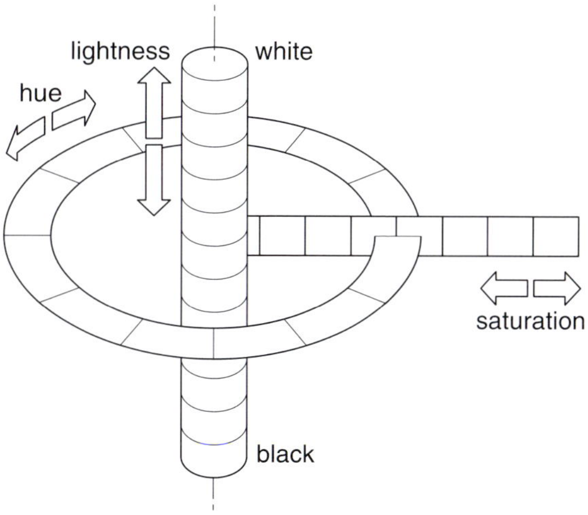

Figure 110 Colour arranged in three dimensions according to the three values of hue, lightness and saturation

The chart in Figure 110 arranges colours (that is, actual samples of the colours) in positions in space. The coordinates of their position act as codes for the colours – up-down for lightness, away from the centre for saturation and around a colour circle for hue. The distances between positions of colours represent their difference – how different they seem to a typical viewer.

4.7.2 Putting colour to work

A designer may want to match colours. This means that a part of a new product is to be coloured the same as the colour of a sample, perhaps on another product or a natural object. A close match might be found in the colours of a chart. With the chart colour as a starting point slight variations in pigments can be tried until an exact match is obtained.

You may wonder why the designer does not just take the nearest colour on the chart. The reason is that most people are very sensitive to colour. If you put two samples of coloured material side by side, even the smallest difference can be noticeable. In fact colours can even be remembered precisely. If you are looking for a familiar product in a shop, you may notice if it is not quite the same colour you are used to. If the colour is slightly paler (lighter and less saturated) than previously, it may suggest the product has been in the shop too long and faded. If the colour is darker and less saturated, this may suggest the product is dirty. If the hue has changed slightly, you might not even recognise the product.

A designer will also want to use colours with appropriate associations and will want to combine colours on a product so that they work together. They might contrast strongly to indicate different parts of the product or be similar to harmonise the appearance. These are difficult decisions for the designer to make and they critically affect how customers will respond. Combining different colours in a single product is made easier by organising colours according to hue, saturation and lightness. However, it is cumbersome to use the whole three-dimensional chart so simplified representations are used.

One way to represent all colours is by using colour circles divided into segments (Figure 111 is a colour circle with twelve segments). Segments of the circle are colours of the same hue. In each segment colours range from weaker, unsaturated colours at the outer edge to stronger, saturated colours towards the centre of the circle.

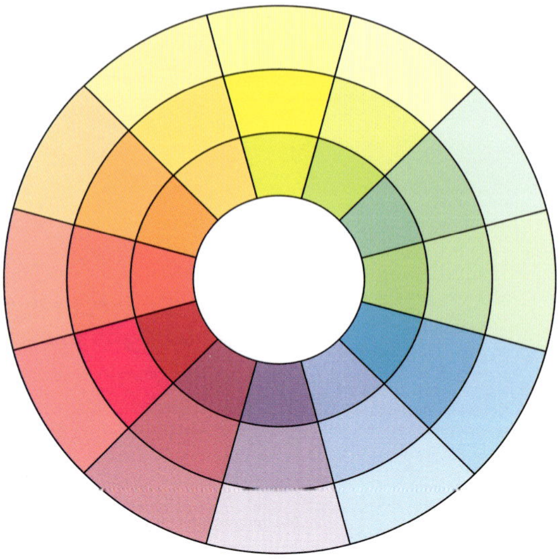

Figure 111 Colour circle – perhaps better described as the hue circle

The colour circle is a common tool when it comes to choosing the colours for different parts of a product. You may be aware from experience of home decorating that there are recommended combinations from the colour circle. Some possible combinations of colours are:

- colours with the same hue but neighbouring saturation (Figure 112a);
- colours with neighbouring saturation and hue (Figure 112b);
- colours with opposite hues on the colour circle (Figure 112c);
- colours of three hues equally spaced around the colour circle (Figure 112d);
- colours of three hues, two closely spaced and the other opposite on the colour circle (Figure 112e);
- two pairs of closely spaced hues on opposite sides of the colour circle (Figure 112f).

Colours help give a product its identity. Choosing colours in a product is a critical decision for a designer. Ensuring that products are manufactured with precisely the chosen colour is equally critical.

SAQ 10

Why is it important to use colours precisely?

W 2

This is a good point to read Sections 7 and 8 of the *Shapes and Space Workbook*, 'Colour as an expressive language' and 'Synthesis, form, style'.

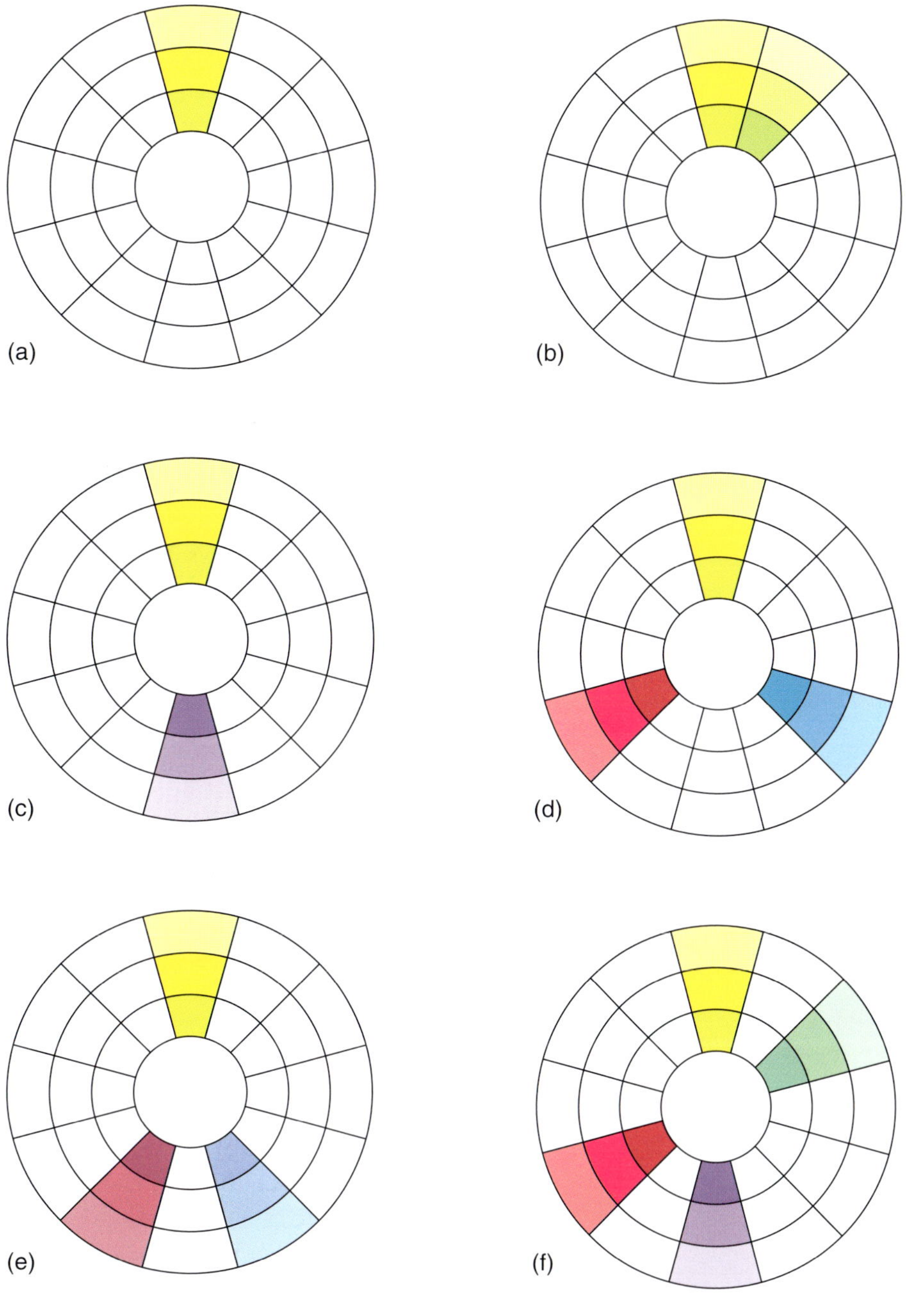

Figure 112 Examples of colour combinations on the colour circle

4.8 Section summary

Component design involves the properties of parts. These properties include shape, size, strength, surface finish and colour. I concentrated on these among the many properties that components possess. I said little about perhaps the most important property of all, which is the material from which a component is made. Block 5 will consider in greater detail the properties of materials.

Key points of Section 4

Cylindrical shapes can have varying diameters of circular cross section.

Constant cross-section shapes include gears and UPVC window frames.

Product outlines can be constructed by joining circles and lines smoothly together.

The perceived smoothness of a curve depends on the curvature changing gradually.

The character of a surface depends on the different curvatures in different directions on the surface.

Material properties have been described that characterise how a component will respond to different loads. A basic set of properties has been introduced: strength, elasticity, toughness, hardness and density.

Surface texture may be created directly during the process of manufacturing a component. Surfaces can also be finished after manufacture by painting, polishing or anodising.

Colour is described by the three values of hue, lightness and saturation. These values specify a colour by its position in a colour chart. Establishing the position of the colour is useful when matching colours or exploring combinations of colours for different parts of a product.

Computer-aided design allows manipulation and visualisation of complex surfaces, which could otherwise only be done by making physical models.

Learning outcomes: 1.4, 1.5, 2.4, 3.1, 3.2 and 3.4

5 Four designers turn concepts into layouts

5.1 Introducing the people, the teams and the objectives

This section presents studies of four designers in professional practice. The four designers and who they worked for are:

1. Hans Petersen, product designer, Cre'active design consultancy;
2. Giles Oliver paired with Richard Owers, architects, Penoyre and Prasad architectural firm;
3. Richard Weeks, engineering designer, Perkins Engines diesel engine manufacturer;
4. Tom Karen, product designer, Ogle design consultancy.

Each designer has a particular way of working. But these designers are not working in isolation because they operate in teams, and they rely on help from specialists outside the team to complete their designs.

In describing the designers and the teams, I will use headings that indicate an objective they aimed for when designing. In sequence, these are 'realising vision', 'meeting the needs of diverse users', 'reducing environmental impacts', and 'exploring configurations'.

Following the themes of Block 4 I will concentrate on the activities of configuration and component design. However, as you have already seen it is impossible to isolate just one stage of designing. The process is iterative. Designing looks forwards to what might be and backwards, for example in the redefinition of design problems. The studies of the work of these expert designers include aspects of the early phase of concept design as well as the later phases of manufacturing and marketing.

Background

Section 5 draws upon a research project conducted as part of the Cambridge-MIT Initiative. This particular project, titled Across Design, has involved a comparative analysis of design processes across different sectors of industry. The four designers were all 'witnesses' in this investigation. In addition the project includes approximately another 25 cases across many sectors from fashion to television, from jet engines to web design. This project ran from 2002 to 2004 between Cambridge University's engineering design centre, computer science department and architecture department; MIT's department of mechanical engineering and architecture department; and the Open University's department of design and innovation.

5.2 Hans Petersen with Cre'active Partnership – realising vision

Hans Petersen is a partner and designer with the Cre'active design consultancy. Cre'active is a company with around 14 designers in two offices, one of which is in Cambridge, England. Petersen is one of four partners at Cre'active. When Petersen was interviewed he was asked about how Cre'active took a concept and developed a layout. He talked about a glucose tester (Figure 113), which was a product the company had designed for a client. What he said about the glucose tester shows some of the stages the design went through until the client was satisfied.

Figure 113 Glucose tester having its internal cartridge inserted

5.2.1 Design consultancy

A design consultancy will work with many different clients developing products for them. These clients may be companies who require anything from whole new designs to detailed advice on one part of a product such as styling. This designer is usually called upon to concentrate on the end users of the product.

The range of products tackled by Cre'active is wide. The range goes from the interiors of trains and buses, to medical laboratory equipment, and from DIY tools to street furniture. Cre'active has a portfolio with few obvious links between products. Some have been made in large numbers, while others are one-off pieces of equipment built for a special purpose. Some were simple with few parts and small, while others were complex with many parts and intricate shapes and connections between parts. Some projects were just concept designs created for clients exploring new ideas. However, the majority are manufactured and sold. Cre'active emphasises its capability to contribute to all aspects of product development – both industrial and engineering design – as well as the detail design required to take a product through testing and manufacture.

The variety of products and the range of capabilities for all stages of product development give a clue to a significant strength of this company. It works with clients from many sectors of industry designing products for many types of end-user. Design consultancies like Cre'active work closely with their clients adding designing skills that complement their clients' capabilities. Perhaps a client does not have its own creative team, or its team may be committed on other projects. Possibly the client may be seeking fresh, new ideas, so it turns to a consultancy.

A design consultancy will work on a project-by-project basis for several clients. It is paid by the clients for the work done. Generally the output of the design work is owned by the client. A design consultancy may create a successful product for a client but it does not benefit directly from this success. However, there are rewards. The client is likely to return to the consultancy for future designs or there may be continuing work on new versions of a successful product. The design consultancy may win prizes and awards for its successful designs, leading to a commercial reputation and therefore work from new clients. However, finding new jobs with new clients still takes Cre'active considerable time and effort.

A typical project undertaken by Cre'active starts with a design being commissioned by the management team at a client company. In 90 per cent of commissions the client is already known to Cre'active. Initial discussion is followed by Cre'active drawing up a programme of work. This programme covers all stages of product development and forms the basis of a contract between client and designer. Early stages have clear deliverables and timescales even though later stages might remain uncertain.

When devising the programme of work the designer needs to be sensitive to potential differences between clients. For example, some clients will require a detailed schedule of work while others are happy to leave planning matters to the consultancy and only need to know key dates and what will be delivered. After a programme is agreed the consultancy will charge agreed fees for design work of various kinds. Drawing up the programme is a necessary preliminary to design work.

5.2.2 Selling a vision

Hans Petersen identifies and sells a vision for the product to the client. He then tries to deliver this as a product that customers can, as he puts it, 'have a love affair with'.

Different customers will be attracted to different features. Some customers, perhaps from Scandinavia, might value strong colour while others from Britain, say, might find a more sober design attractive.

Notice the distinction here between client and customer. The client is the company commissioning the design, a customer is someone who will buy or use the product.

5.2.3 Realising a vision – a glucose tester

The early client meetings define concept. Cre'active will produce physical models of possible configurations for clients to handle. For example, different possibilities for the glucose tester shown in Figure 113 were modelled in a white plastic so that clients could handle alternative designs without the distraction of colour and applied graphics.

A glucose tester is used by diabetics to test the levels of glucose in their blood. The round shape that was devised fits easily in the palm of the hand and is about 65 mm across. Inside it is a circular cartridge with small, encapsulated drops of a pharmaceutical that test for blood glucose (Figure 113). With the device closed, diabetics test themselves by pressing one end of the tester to their skin and pushing a slider on the top with their thumb. This movement conducts the test. The slider then springs back to a home position, uncovering a small display with a record of the result.

Creating a physical model for evaluating the overall shape as well as the arrangement of the different parts was particularly important in this case. The tester was to be hand-held by diabetics who may have impairment of manual dexterity. Notice the early concern with usability in developing the design. The client also needed a physical model because two of the main issues being explored were the kind of cartridge to use and how the test was to be done. The circular cartridge looked like a good solution but before this idea could be developed further the client needed convincing. Therefore the physical model was constructed, which could be handed round at the meetings.

For the glucose tester two features of the layout that had to be configured were a shape suitable for holding and good access to batteries. The surface needed large, bold features so that the tester did not slip in a user's hand while it was being operated. The finished product had a circular cartridge of ten separate tests. The adoption of this circular cartridge rather than a card or strip was also a major part of the layout. A third challenge was designing how the device was used. A slide button was incorporated with a click-click forward and back movement.

The negotiation with clients on requirements, data gathering about patterns of use, development of concept and initial layout took about eight weeks. The time devoted to negotiation varies considerably from project to project. A team of three or four designers worked on the project – this is typical for this kind of product. At the early stages members of the team offered alternative designs but as negotiation with the client proceeded one concept was selected and developed to layout.

Throughout the process of going from first client briefing to concept and then to layout, experiments and models were applied to understand the user needs and how they would be met by the proposed designs. In parallel, details of appearance and feel were developed with a client that relied strongly on the design team for suggestions. However, the client needed convincing that the design proposals really met their requirements and the needs of the end users of the product. Colour, shape and texture had to convey the right image of the product and also the client's brand.

A surprising feature of many of Cre'active's projects is the long time from the first meeting to the product appearing on the market. This is not such a problem where products can have a long life, in the sense that they stay in production or use for many years. However, most companies, including design consultancies, are continually trying to reduce lead-time – the time from conception to market. The glucose tester required about three years to bring to market but may still be made and used in 10 to 15 years time, though it may be updated in that period. Indeed, Cre'active hopes to get a contract to undertake any design modifications and updating.

As well as healthcare Cre'active does design work in the transportation sector. The company has had contracts with London Underground and various train operators around the world. Trains have taken five years from design brief to entering service. They can then remain in service for over twenty years before a major refit.

Another area of Cre'active's design work is street furniture. An example is the railings in a pedestrian area near the Baltic art gallery and Gateshead-Newcastle Millennium bridge, which are shown in Figure 114.

Figure 114 Railings near the Baltic art gallery and Gateshead-Newcastle Millennium bridge

5.2.4 After layout

So what happens in the considerable time after a layout is confirmed until a product is launched? Apart from product refinements, the design is analysed to ensure it will perform as required, materials are selected and prototypes are manufactured and tested. Approximate costings are made. At this stage, the client takes the decision whether to go with the product or not. The next steps, to manufacture, distribute and market the product, are costly. Of course expenditure on design has already been significant but is small beside these costs.

A consultancy like Cre'active may be closely involved in these decisions, drawing up costs as well as making and testing prototypes. Cre'active finds that not all its designs proceed to market. These assessments might prove the design has faults or the risk of going to market is too great. It is even possible, because of the time lag between concept and launch, that the market conditions have changed.

To summarise, Hans Petersen provides a vision for the product, introducing an intrigue factor into designs that will entice a customer. The majority of Petersen's design work is in realising the vision – transforming concept to layout. Petersen wants his designs to be noticed above others – hence the intrigue factor. He wants a Cre'active design to be the first one that a potential customer reaches out to try.

Exercise 21 Steps in designing

From the description above list the main steps in designing the glucose tester. For each step identify the main type of activity taking place using one of these: briefing, concept, layout – parts, layout – arrangement, layout – connections, layout – form. So the first might be, 'First meeting with client (briefing)'.

Discussion

From the description I have identified the following steps.

1. First meeting with client (briefing).
2. Devise programme of work (briefing).
3. Analyse need and constraints (concept).
4. Generate several competing concepts (concept).
5. Decide on the main parts – circular cartridge, sliding button, circular shape with natural features to hold steady in the palm of a user's hand (layout – parts).
6. Configure the main parts (layout – arrangement).
7. Determine details of catches, release mechanisms and hinges for opening the tester to remove empty cartridge and insert new. This is about the connections between the main parts (layout – connections).
8. Finalise component shapes for holding and operating (layout – form).
9. Determine texture and feel of surface of body and button, and the colour (layout – form).

5.3 Oliver and Owers with Penoyre and Prasad – meeting the needs of diverse users

Giles Oliver and Richard Owers are architects who worked on a design for an urban housing scheme. Oliver and Owers were in the project team set up by the architectural company Penoyre and Prasad. The project team had the task of developing a plan for a large urban site in Gravesend near London.

Penoyre and Prasad is a firm of architects established in 1988. Its projects range in scope from individual buildings to urban master plans. Projects have included education, healthcare, sports, community buildings, housing, concert halls, cultural centres and city blocks. The company says to clients that it will 'provide design quality without compromising programme and budget'.

Oliver and Owers worked with a range of participants, from councils and resident groups to developers and potential occupants in addition to the direct client that commissioned the design work and paid for it. The main role of the architect here was to present possible schemes through drawings and models to all these stakeholders in order to facilitate informed negotiation among them.

5.3.1 Meeting diverse standards

The key motivator for Oliver and Owers was *quality*. This meant the designer wanted to meet the standards expected by several clients, users, and the buyers or renters of property. In practice these diverse standards were not all met and tradeoffs occurred. The architects explained to stakeholders why tradeoffs were made and what would be the likely impacts.

As well as the architects, the design for the site involved interested parties such as planners, council officials, residents and developers. All parties had an intense interest in the design. For residents it is the environment in which they will live, for planners and councillors the development will express the culture and vision of their town. They have a duty to residents and are often the channel for external funding, which in this case was through regional regeneration budgets from national government and the European Union. Developers, usually private companies, buy land, pay for the design and build, and sell on to individual buyers. Their interest may be short term, lasting until they receive revenues from sales. They will be keen to make sales as quickly as possible in order to recover their investment.

People have strong connections with their houses, streets and towns. User needs can be demanding, diverse and maybe contradictory. For example, an individual's need for amenity may conflict with their need for convenience and transport. A person's immediate built environment rarely changes dramatically, with houses, shops and places of work forming the framework for their lives. Buildings are not like products, they cannot be discarded or easily changed for a new model. This is especially true of the general appearance and character of towns and cities.

5.3.2 Phase one, development – phase two, building

There were two clear phases to this project. First, the development led to a plan and proposals for the different buildings. This was a good example of turning a concept into a layout. Second, the building phase was where an architect took the layout and worked on the details of the design to meet the needs of both the developer and future residents.

The architectural firm of Penoyre and Prasad was appointed as master planners for the development phase. The site for redevelopment was a car park on land where housing had been demolished 30 years before. Before that, in 1826, the site was a brickfield (see Figure 115).

1826 1935

Figure 115 Gravesend site 1826, 1935

The architects at the second phase may, or may not, be the same as those at the first phase. Indeed one reason for the firm of Penoyre and Prasad to engage in the development plan is to be well placed to be the architect during the building phase. Another reason is to be able to ensure high levels of quality throughout. If the architect had just been engaged for the second phase, many of the important decisions would have already been taken.

The way Oliver and Owers talked about quality may be different to the way you use the word. You may think of a quality product as being carefully designed, manufactured with great attention to detail and possibly with no real consideration for cost. This is not quite what Oliver and Owers meant.

What Oliver and Owers were getting at was that to achieve quality they would have to maximise the desirable features of a building even though there were severe constraints. For example, in a new apartment, an expensive floor finish may be included but at the expense of a more efficient heating boiler.

The assessment of quality here is complex. Different participants may see the floor-versus-boiler decision differently. The developer may see the floor finish as more desirable than the boiler because it will promote quick sales of the apartments. Some buyers will not consider floor finish adding to quality – they may want to lay their own – and be more concerned with the running costs of the boiler. Others may have easy resale of the property as a priority and take a view that nice floors will be a selling point while an efficient boiler will not.

The architects saw their role as presenting new possibilities to the various stakeholders. They modelled different possible layouts for the development. They did this with drawings, computer models and scale models. It is around these proposals that all the parties negotiated. The stakeholders expressed likes and dislikes, estimated costs, and assessed visual appearance by reference to the models the architects presented.

This process of negotiation went through many stages. Different parties at different stages worked with different kinds of models. At early stages land use and integration with surroundings was of primary concern. At later stages, perhaps in negotiations with developers, the details of house design – sizes, specifications of materials and costs – were of primary concern.

Architects on the development scheme in Gravesend used design quality indicators (DQI) to assess designs. These indicators were divided into three types. *Functionality* concerned how a design met specific end-user needs. *Build quality* was about the how the design performed technically. Performance of heating and ventilation systems, for example, would be assessed according to efficiency, reliability and maintenance schedules. The *impact* of the design concerned visual character, environmental impact of the use of certain materials (in their production and disposal during the life cycle of the building), the comfort of occupants and the extent to which the building would fit with the existing or proposed urban and social fabric. Design quality indicators offered a framework to assess a building design and made all parties to a design aware of tradeoffs that would have to be made.

Table 1 Design quality indicator fields

Functionality	Build quality	Impact
use	performance	form and materials
access	engineering systems	internal environment
space	construction	urban and social integration
		character and innovation

5.3.3 Different models tailored to different needs

Designers often facilitate communication among parties. They can interpret and externalise intentions and wishes of parties in a negotiation – often by using their models. Their conjecture can stimulate discussion. The use of various models by architects can be valuable as the development plan moves through its various stages.

At the start of design a computer-aided design (CAD) model of the site was constructed (see Figure 116). This showed the surrounding buildings as grey blocks. Possible development schemes were then inserted in colour. This allowed the architects to present different schemes in a realistic bird's-eye view so that planners and residents were able to visualise proposals.

Figure 116 CAD model and four sketches of the development sequence

Oliver and Owers found that these CAD models took considerable time to prepare prior to meetings. Speedier tools for visualisation, such as sketching, were used at meetings to help the parties share their ideas on how the proposal might be developed. The architects made extensive use of sketches to show the main areas, access, views and the possible sequential development of the site. These quick sketches were done in meetings with planners and residents. In the process of negotiation many sketches were used alongside the CAD representations that were created before the meetings. Other representations used during development included proposed layouts for the whole site (Figure 117).

Figure 117 Proposed layout for the whole site

The proposal was to develop the whole site in a sequence of four building projects. The four sketches you see in Figure 116 refer to these four projects. The design for the first project in the sequence developed one end of the site. Oliver and Owers and the team then developed this first project in more detail.

This first project consisted of a block of housing around a courtyard, with a tower – a high-density development of 105 flats above a commercial area. Several representations were used, including CAD renderings of the whole scheme (Figure 118a), scaled three-dimensional models in card and wood, and perspective views of movable glazing (Figure 118b) for apartments in the courtyard. Layouts for apartments were also provided (Figure 119). The arrangements of glazing were presented merely as concept sketches rather than exact realisations of what any particular apartment would look like.

It is instructive to note how the individual pictures do not always match. The architects were using their representations to explore possibilities. They had not yet reached the point at which they needed to make all the drawings exact. However, later, all the drawings had to be exact and consistent before building started.

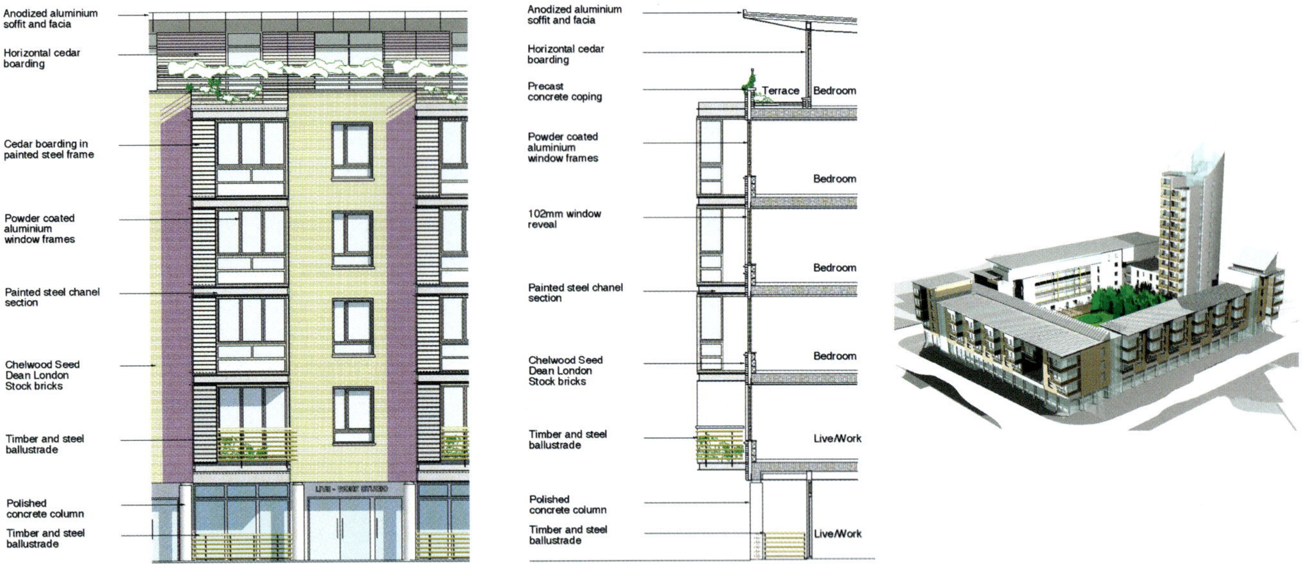

(a) Façade, section and perspective views

(b) Internal arrangement of windows and screens

Figure 118 Apartments

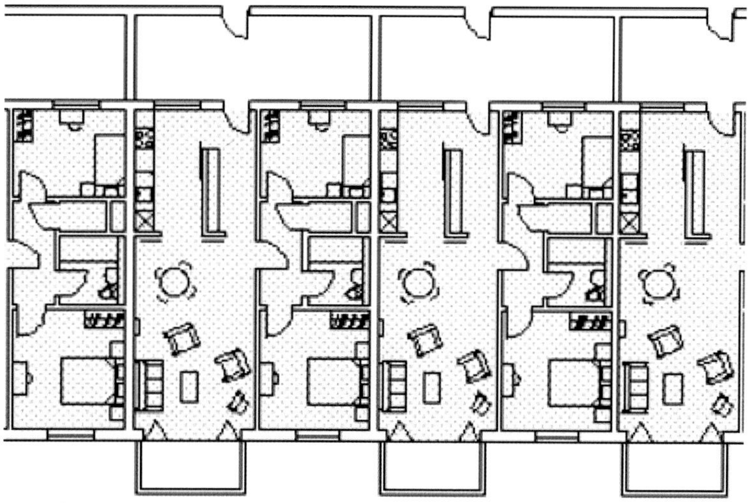

Figure 119 Apartment layouts (detail taken from plan)

These proposals for apartment design and layout were not made with reference to particular developers. The architect's client was still Gravesham Borough Council Regeneration Services, which commissioned the master plan. However, the council needed to take the design to sufficient detail to start negotiations with a developer. Preliminary decisions on the types of apartment and the types of buyer needed to be made. In this case the scheme included houses to rent through a housing association and houses for sale.

After the development phase, the building phase brought a developer into the design process. The developer came with specialist knowledge of the requirements of buyers. It turned out these were not quite the same as those of the council or the architects. As the design proceeded, more user needs, and more precisely defined needs, influenced decisions on the design. One example of this is that the developer was keen to have many apartments with two bedrooms and two bathrooms. This was not to cater for families but for owners who wished to rent out a room to offset a mortgage, but did not wish to share a bathroom.

There are two points to make before you do the next exercise: one about designers buying specialist help from other companies, and one about environmental impact.

First, the architects in the Gravesend project worked in collaboration both with the client and with a number of consultants. Landscape architecture, CAD models and cost estimates were all bought in by the designer as needed.

There is a generalisation to make here. As designers move towards detail design they may need to seek specialist advice. This is not just the case for large projects such as buildings but is true for almost all design projects. For example, analysis of manufacturing processes and detailed costings of a product design (or components going into a product) will need to be supplied by a specialist manufacturer.

Second, the architects in the Gravesend project were particularly concerned with environmental impact. This included taking care over the materials used to reduce energy consumption, dwelling density (there are 270 habitable rooms or 108 dwellings per hectare in this proposal, which is high density) and maintenance requirements for the life of the buildings. The architects engaged environmental consultants as well as using their own computer models to calculate energy consumption and costs for different options. For example, if the apartments were well insulated then heating costs would be low, so investing in super-efficient boilers would not be appropriate.

In housing schemes where there are concerns for energy saving, any decision will have many possible combinations of options on heat from boilers, solar, fuel cell or combined heat and power, and many options on types of insulation. Computer models are good at evaluating these combinations of options.

Exercise 22 Property developing

Examine the layouts of apartments in the first stage of the Gravesend development (Figure 119). Sketch one suggestion for incorporating another toilet with a shower or bath in the two-bedroom flats. Roughly these flats comprise two long rectangular areas, one containing bedrooms and bathroom and the other the living room and kitchen. You are not being asked to solve this problem, just to make an outline suggestion and draw a sketch.

Discussion

It seems that keeping two 'separate' rectangular areas of living and sleeping imposes a severe constraint. I thought I would start by removing this constraint and using the whole rectangular space. I sketched out a possible layout of four main spaces – living, bedroom 1 and bathroom 1, bedroom 2, and a second toilet and shower (Figure 120). The second toilet and the shower are to be used by the second occupant and visitors. Don't forget the entrance. You might find the specimen bathroom layouts in Section 1 useful.

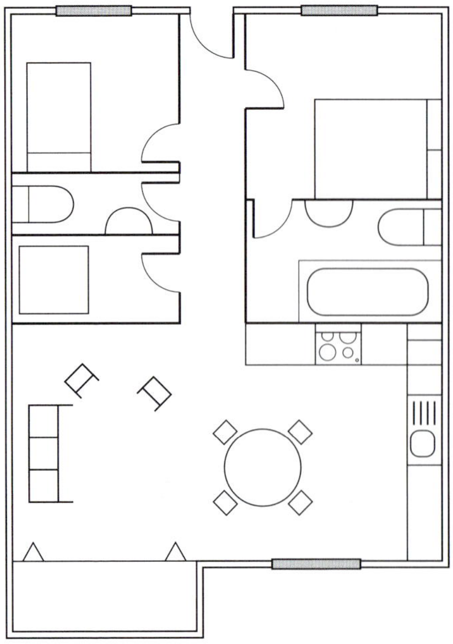

Figure 120 Two-bedroom apartment layout

SAQ 11

List the types of representation used in designing the Gravesend development plan.

5.4 Richard Weeks with Perkins Engines – reducing environmental impacts

At the time Richard Weeks was interviewed for this study he had a senior role in product development of engine design at Perkins Engines, which has headquarters in Peterborough, England. Richard Weeks is a designer at Perkins and his responsibilities covered all stages of design from the concept to detailed layout, and then onwards to ensure manufacturers and suppliers meet design requirements and ensure customers receive the designs agreed.

5.4.1 Engineering design, client needs and environmental regulations

Weeks was mainly concerned with function and performance. He was part of a large team developing a range of engines for use in products such as electricity generators and tractors. Key concerns for this designer were performance and cost, but the critical factor in this field of design was new regulations for exhaust emissions. The engines had to meet these if they were to be approved for use in the main world markets.

Figure 121 Examples of Perkins diesel engines

Perkins Engines is a world-leading manufacturer of diesel and gas non-road engines (see Figure 121). The company does not produce engines suitable for on-road use, so the engines are not going into cars, buses, or lorries. Engines range in power from 4–2000 kilowatts (5–2600 horsepower) and their physical sizes range from knee height up to engines with cylinders you can stand in.

Initial engine concept is based on a specification from a particular customer wanting engines for a particular product being designed.

Perkins designs and manufactures engines that broadly consist of three main parts: the engine, the fuel delivery and the electronics to control operation. The engine itself has several parts such as filters and pumps that are arranged on the exterior of the engine. The physical configuration of the engine is carefully designed. One reason is that many customers need engines to fit a specific packaging space. In the case of military vehicles the packaging space for engines is limited and tightly constrained. Military engines are designed specially for each vehicle type.

Regulations on energy efficiency and reducing emissions from energy production have a significant effect on designs. Exhaust emissions from engines are now regulated strictly.

A strict timetable of emission and efficiency regulations is in place, which Perkins' designers have to meet. Perkins sells its diesel engines to construction plant and electricity generator manufacturers that naturally demand compliance with these regulations. However, unlike many market requirements, regulations are well defined many years ahead.

Table 2 Non-road diesel engine emission standards of the US Environmental Protection Agency

Engine power (kilowatts)	Tier 2			Tier 3		
	date	NO_x + HC (g/kW hour)	PM (g/kW hour)	date	NO_x + HC (g/kW hour)	PM (g/kW hour)
< 8	2005	7.5	0.8	2009*	5.0*	0.3*
8–19	2005	7.5	0.8	2009*	5.0*	0.3*
19–37	2004	7.5	0.6	2008*	4.7*	0.17*
37–75	2004	7.5	0.4	2008	4.7	0.13*
75–130	2003	6.6	0.3	2007	4.0	0.13*
130–225	2003	6.6	0.2	2006	4.0	0.13*

The figures marked (*) are provisional in 2004. Legislation harmonized with the EPA standards has been introduced by the European Commission.

Looking at Table 2 the first column indicates the power output of the engine. Different powers have different schedules in a rolling programme of regulating emissions. Tractor and truck engines are in the middle bands from 19–130 kW. These numbers represent how much energy the engine can produce in a unit of time. The third column specifies the targets for emissions of nitrous and nitric oxides (NO_X) and hydrocarbons (HC) in terms of grams per kilowatt hour. So an engine with an output power of 50 kW running for an hour produces energy of 50 kilowatt hours so has limits on emissions of NO_X + HC of 375 grams (7.5 x 50) in the hour. This engine, run over a (short running) life of about 5000 hours, gives NO_X + HC emissions of 1.875 tonnes (5000 x 375 grams) over its life. The fourth column considers particulate matter (PM). The same engine will be limited to producing about 200 kilograms over its 5000 hour working life.

Notice the regulations have tiers. The engines related to tier 1 (not shown) are incorporated in engines now (2004). Tier 2 is the immediate future with absolute and agreed limits that all designers should work to. Tier 3 is expressing feasible targets and some remain to be finally agreed.

On the one hand, a design team at Perkins with around 40 engineers working on particular engines was working to customer specifications such as power output, maintenance intervals (for example oil and filter change), and duty cycles (proportion of operating time at different loads). On the other hand, all the design teams had to ensure that emission regulations were met across the range of engines.

5.4.2 How much is new?

A significant proportion of design at Perkins is about realising existing concepts in new layouts for different customers. The product is mature in two senses. First, the level of newness in an engine design is small. Second, competitive advantage over other manufacturers is achieved by tiny differences in engine design.

It might be surprising therefore that the concept of newness was used a lot by designers at Perkins. If parts had been used before on another engine, they would have been tested and proved in service.

Provided the conditions are similar designers knew how these parts would behave. These parts were not new. A part was new if that exact part had not been used before. If the part had been modified, however slightly, in a feature that might affect how it worked, then it was a new part.

A guide to newness was set at around 25 per cent in terms of number of new parts. In many cases the newness in a part may just have been a change of size or a slight change in shape. The measure of newness is used to indicate the risk involved. An estimate of 40 per cent newness indicated a very high risk that needed to be managed carefully. How was this risk managed?

For a design with 40 per cent newness risk, special care was given to each decision. Just one wrong decision was potentially disastrous and could have serious knock-on effects as other decisions were made. The design team analysed and tested new designs with new parts. This was costly but it reduced the risks of an engine failing in service. Perhaps failure is too strong a term here. It was more likely the engine would not perform quite as well as a competitor's engine. For Perkins, with its mature products, this would count as a failure.

5.4.3 Modelling methods

Analysis of a proposed design was first made on computer models. These included three-dimensional models of the layout of all the parts and other models of how the engine performed in terms of fluid flows, electronics and mechanics. An example of analysis was ensuring the strengths of different parts of the engine could bear the loads as the engine ran. The shapes of all the different parts had to fit together and be accessible. These were just the kinds of detail that were defined in the fourth stage of layout design. Computational analysis and testing made sure the designs performed satisfactorily.

Perkins always used CAD models so that ambiguity and misunderstandings between designers, suppliers and customers were reduced as much as possible. One of the problems with informal communications such as sketches was they could be understood by different people in different ways. This may not have been a problem within a design team but could cause confusion outside the team.

What kinds of models were used by Perkins at the earlier stages when the design team was exploring configurations? In common with product designers and architects, physical models were used. In Perkins' case these even included modified pictures of existing engines to convey an impression. Sometimes parts from existing engines were put together, not as a working engine but as a mock up. This may have involved gluing and welding parts together to help visualise the overall configuration.

However, once it was decided to develop a particular engine (based on a business case presented and argued within Perkins itself) designers created detailed engineering models of parts, shapes, sizes and configurations. These details were needed to analyse performance of possible designs.

The designs for diesel engines created by Perkins are manufactured by the company. The industry is very competitive and the manufacture of engines should cost as little as possible. To do this it is necessary to ensure the layout avoids complicated assembly where possible. The designer may need to think ahead to manufacture and therefore take into account the sequence of assembly operations, accessibility and the fixing operations. Also, the designer needs to consider individual components and how they will be manufactured. How these important decisions on design for manufacture and design for assembly affect the success of products will be examined later in the course.

5.5 Tom Karen with Ogle – exploring configurations

Tom Karen is a designer who is well known for several individual and immediately recognisable products over the past thirty years. As managing director of the design firm Ogle, he had significant input to designs such as the Bush TR130 radio, the Bond Bug car, the Reliant Robin car and the Raleigh Chopper bicycle. Ogle is a consultancy doing design projects for many clients.

5.5.1 Karen's background

Karen's designs described below are from around 1970. He has had many design successes since. However, the historical examples will allow me to take a more detached view of a designer at work.

At the time of writing Karen was retired but was still designing. His studio is now at his home and is a place full of objects he has made, together with samples of materials, colours and shapes.

I thought Karen was worth interviewing because his work illustrates how changing the configuration of parts in a product can be a key driver in new designs. He made strong design statements for clients and met user needs.

Some of his designs are rather idiosyncratic. Many have been successful and the Bush TR130 radio is still sold as a reproduction. They might not come top in a poll of favourite designs but they have a life and vibrancy about them, which has made them endure as design classics. They are curious, they intrigue, and some have been phenomenally successful.

(a) Reliant Scimitar GTE

(b) The predecessor of the GTE, the Scimitar GT

Figure 122

Around the late 1960s, Karen worked extensively for Reliant Motors. The company's Scimitar GTE was a successful piece of innovative styling launched in 1968 as an addition to an existing range of Scimitar cars. One predecessor, the Scimitar GT, not designed by Karen, had a three-box, two-seater configuration. That is, box 1 was the engine compartment at the front, box 2 was the passenger compartment in the middle and box 3 was the boot extending at the back. With the Scimitar GTE, Karen introduced a two-box design with a hatchback. A long roof and rising waistline at the rear gave a compact and sporty appearance as well as improving its aerodynamics. It was one of the first four-seater sports saloons. The Scimitar GTE was a successful example of car styling, which took advantage of a demand in the market for a four-seater sports saloon car. The new design was driven by a change in the configuration from three-box to two-box. The Scimitar GTE had a glass-fibre body with steel chassis, in common with all other Reliant cars.

To put this design into context, I will briefly describe another of Karen's designs for Reliant in the late 1960s, the Bond Bug. The Bond Bug was not successful in market terms, but it was an interesting vehicle. This was a wedge-shaped three-wheeler in a long tradition of small, lightweight vehicles that included the so-called bubble cars. The innovative design was a one-box car that was simple to assemble. Only 3000 Bond Bugs were made because the whole market for three-wheel cars was small, in spite of low running costs. Competitors such as the Mini offered more space, comfort and safety.

5.5.2 Exploring configurations

Exploring new configurations led to some of Karen's commercially successful designs. Remember that designers like Tom Karen are contracted by clients to produce designs, they generally have no other commercial interest in the product. If their designs are successful in the market, they hope to be rewarded with further design work from the client and others.

Figure 123 Bush TR130 radio (1964)

The Bush TR130 radio originally launched in 1966 and relaunched in 2000 as a reproduction has sold in large numbers (Figure 123). Transistor radios had followed older and larger valve sets with dials and tuning knobs on the front of the radio. The larger size allowed both controls and the loudspeaker at the front. The TR130, although not the first, located the controls and tuning indicators on the top of the radio. Push buttons were used for selecting wavebands. Karen was exploring possibilities for the configuration – controls on top, push buttons and the speaker grill covering nearly the whole of the front of the radio. However, success in this case comes not only from configuration but also from the opportunities the configuration offers. Karen identified as key features the curved 'magnifying glass' over the tuning scale and the way the user could pick up the Bush TR130 like a book rather than using the floppy handle. The brown covering around the grill was easy to grip. The magnifying glass feature is about the appearance of the radio and the covering about its feel. Both appearance and feel are critical to a product's success in the market.

As you have seen in his car designs, Karen explores the limits of new configurations, giving an individual and strong identity to the design. The silver grill of the Bush TR130 dominates the front, the hardness of the grill is set off by the softer texture of the surrounding fabric, the controls are set back from the top edge – these were all decisions made about the configuration. In some sense he pushed the configuration towards the edge of what was possible, yielding a design that remains interesting and attractive to customers over thirty years after it was designed.

Another successful design by Karen that sold in large numbers was the Raleigh Chopper bike launched in 1969 (Figure 124c) and relaunched in 2004. Again this shows how he subtly changed configuration. It is based on earlier US bike designs, starting with the Schwinn Stingray in 1963 (Figure 124a). Karen was engaged by Raleigh to design a bike for the UK market based on the Raleigh Rodeo (Figure 124b), which it had been selling in the US in 1966.

(a) Schwinn Stingray (1963) **(b) Raleigh Rodeo (1966)** **(c) Raleigh Chopper (1969)**

Figure 124 Two configurations of bicycle that preceded the Raleigh Chopper

Exercise 23 On your bike

(a) Compare the Stingray and the Rodeo. Identify similarities in configuration and shapes of components.

(b) Now compare the Rodeo and the Chopper. Identify differences in configuration and shapes of components.

Discussion

(a) The configuration of the Stingray and Rodeo is similar. The shapes used in the Stingray and Rodeo are similar except for larger wheels on the Rodeo and a curved seat on the Stingray, a protected feature of this design.

(b) The Chopper has a different arrangement of straight tubes, rather than curving tubes, forming the frame. The central gear lever of the Rodeo and Chopper are similar but the seat of the Chopper is different because it rises higher at the back. The wheels of the Chopper are smaller than the Rodeo, and the Chopper's front wheel is smaller than the back. Generally the whole appearance of the Rodeo is that of a traditional bicycle. The Chopper, with some clever changes to configuration and shape, is a new design.

Key points of Section 5

Designers work with many different parties during the course of a project, including clients, users and future exploiters of the design.

Designers provide a range of representations to facilitate communication and negotiation among parties – sketches, models, 3D computer-aided design views, computer analysis and drawings.

Close attention to user needs is vital, but a designer must also have a vision for the product. This vision needs to be communicated to clients and customers.

Design is tightly controlled by regulations, especially in the area of safety, and increasingly in environmental impact.

For mature products such as engines, managing risk is part of designing. Each new product incorporates only a few new components (components that have not been tried and tested in a previous product).

Exploring alternative configurations and changing the shape of parts can help create new and successful products.

Learning outcomes: 1.2, 2.5, 3.2

W 2

If you haven't already done so, now would be a good time to read the last section of the *Shapes and Space Workbook*.

Answers to self-assessment questions

SAQ 1

Identify three pieces of information that a layout contains.

1 The sizes and shapes of all the parts (including a list of parts).

2 The arrangement of the parts in a configuration.

3 The connections between parts.

SAQ 2

What are the two essential activities for an iterative process like design?

The first group of activities concern evaluating design proposals against requirements. The second group of activities concerns generating possibilities. For generation, sometimes rules like the necklace of furniture around a room may be useful. This is because the proposals generated are likely to be feasible designs.

SAQ 3

Think about the life cycle of a washing machine. Identify three different groups of people who might be affected by raw material extraction, manufacture, distribution, use and disposal.

You might like to distinguish groups according to geographical location relative to the first user. There are many groups. Some people will be directly affected by raw material extraction, and they may be in a different country; so different national groups may be affected. Other groups may be affected indirectly such as from the energy generated or from pollution during extraction.

Within a country, a regional population will be affected by requirements for raw materials and treatments in the manufacture of washing machines. A regional population will be affected by the processes associated with use – for example water treatment and water supply.

Regional and national populations are affected by transporting the machine during distribution to customers, as well as in transporting parts from suppliers to be assembled in the final product.

Local populations will be affected by disposal of those parts and materials that cannot be reused or recycled.

SAQ 4

Identify two examples of structural design in which triangles of parts are used extensively.

Familiar examples are cranes and roofs. The roof of the covered courtyard at the British Museum is a good example. However, perhaps the example you are most familiar with already from this course is a bicycle frame. The Strida configurations are essentially triangles.

SAQ 5

For the two patterns of lozenges shown in Figure 44, find three symmetries for the first pattern and one symmetry for the second.

If the pattern is reflected in the lines a or b, it remains the same. These are two reflection symmetries. Rotation through 180 degrees about central point O also leaves the pattern the same – this is the rotation symmetry.

The second pattern is the same geometry as the first but is now coloured. Only some of the previous symmetries will still hold. Reflecting in lines a or b changes the pattern as the red and black lozenges are interchanged. However, the 180 degree rotation keeps the pattern exactly the same so is the only symmetry of the coloured pattern.

SAQ 6

Figure 65 shows room dimensions. Around the perimeter are letters standing for the distances between walls. Work out the distances a, b, p, q, r, s from the dimensions of the rooms shown on the plan. for example, a=4 and p=3.

a=4, b=4, p=3, q=6, r=1, s=5

SAQ 7

Give an example, not referred to above, of cylindrical shapes used in domestic items whose cross sections change.

There are many examples you might think of. One example is a wine glass. The sections of the wine glass taken across the stem and bowl are all circles but at the bowl of the wine glass the sections are rings – one circle inside another. Other examples are a rolling pin, a flowerpot or a cylindrical pen.

SAQ 8

Have another look at the soft drinks can that you examined earlier in this section (Figure 82). Every part of the surface is curved. Which parts curve in more than one direction? Identify the parts of the surface with the highest curvature. Try to answer these questions before looking at Figure 94.

Referring to the section of the can shown in Figure 94, the straight sidewalls are curved in one direction. The convex shapes at the top and bottom of the sides are curved around the can and up and down. The base of the can is also curved in two directions. These curved shapes contribute to the strength of the can, which is made from thin aluminium sheet. The highest curvature occurs at the shoulders and the base of the can.

SAQ 9

Look at the picture of the iPod in Figure 102. How would you specify the outline of this front view?

You would need to specify the overall length and width and give the radius of the corners. As each corner is 90 degrees – ignoring the rounding – this would describe the entire outline.

However, there are other outlines, as when viewed from the side. The corners are complex three-dimensional shapes. Their exact shape is central to the feel and appearance of this product. However, the question just asked you to specify the outline in the view from the front.

SAQ 10

Why is it important to use colours precisely?

Colour has strong associations. People can discriminate between colours that are slightly different. So if a customer is looking for a familiar colour in a product or the designer wants to achieve a special association, a precise colour is necessary. It is important to ensure the desired colour is reproduced and maintained in the final product. To do this the colour is matched to a precise reference using a colour system such as Munsell or Pantone.

SAQ 11

List the types of representation used in designing the Gravesend development plan.

The description of the project mentioned the following: sketches, 3D CAD, site plan, plans and elevations, concept sketches of movable internal glazing, energy models, details of glazing.

References and further reading

Alexander, C. (1964) *Notes on the Synthesis of Form*, Cambridge, Mass., Harvard University Press.

Antonsson, E. and Cagan, J. (eds) (2001) *Formal Design Synthesis in Design*, Cambridge, Cambridge University Press.

Ashby, M. and Johnson, K. (2002) *Materials and Design*, London, Butterworth–Heinemann.

Birren, F. (1969) *Principles of Color*, New York, Van Nostrand Reinhold.

Bucciarelli, L. (1994) *Designing Engineers*, Cambridge, Mass., MIT Press.

Critchlow, K. (1970) *Order in Space: a design source book*, London, Viking.

Flurscheim, C.H. (ed) (1983) *Industrial Design in Engineering*, London, The Design Council.

Gordon, J. E. (1991) *The New Science of Strong Materials*, Penguin, London.

Henderson, K. (1998) *On Line and On Page – visual representations, visual culture, and computer graphics in design engineering*, Cambridge, Mass., MIT Press.

Jordan, P. (2000) *Designing Pleasurable Products: an introduction to the new human factors*, London, Taylor & Francis.

Judd, D.B. and Wyszecki, G. (1975) *Colour in Business, Science and Industry*, New York, John Wiley.

Knight, T. (1994) *Transformations in Design*, Cambridge, Cambridge University Press.

March, L. and Steadman, P. (1971) *The Geometry of Environment*, London, RIBA.

Otto, K. and Wood, K. (2001) *Product Design*, Prentice Hall, New Jersey.

Pahl, G. and Beitz, W. (1996) in Wallace, K.M. (ed) *Engineering Design – a systematic approach*, second edn, London, Springer-Verlag.

Simon, H. (1969) *The Sciences of the Artificial*, Cambridge, Mass., MIT Press.

Stevens, P. S. (1981) *Handbook of Regular Patterns: an introduction to symmetry in two dimensions*, Cambridge, Mass., MIT Press.

Suh, N. P. (2001) *Axiomatic Design – advances and applications*, New York, Oxford University Press.

Sullivan, L. (1956) *The Autobiography of an Idea*, New York, Dover. Published earlier in 1924.

Thompson, D. W. (1917) *On Growth and Form*, Cambridge, Cambridge University Press.

Ulrich, K. and Eppinger, S. (2004) *Product Design and Development*, third edn, New York, McGraw-Hill.

Wankel, F. (1965) *Rotary Piston Engines*, London, Illiffe.

Acknowledgements

Grateful acknowledgement is made to the following sources for permission to reproduce materials within this book.

Illustrations

Figures 1, 2 and 55: MAS Design Products Ltd; Figures 5 and 30: Copyright Black & Decker Inc. 2001; Figure 7: Lionel March and Judith Sheine (eds.), *RM Schindler Composition and Construction*, Academy Editions, Academy Group Ltd.; Figure 12: Mary Quant; Figures 18 and 19: British Standards Institution; Figure 25(a): David Mellor; Figure 51: Alessi Showroom, 22 Brook Street, London W1K 5DF; Figure 52: Dyson; Figure 54: Crown copyright, reproduced with permission from the Controller HMSO. Figure 63: © 1963 Detsche Verlags-Anstalt, Abterlung Fachvolag, Stuttgart; Figure 71: Source: Commission Directive 96/60/EC of 19 September 1996 implementing Council Directive 92/75/EEC with regard to energy labelling of household combine washer-driers, Official Journal L 266, 18/10/1996, P. 0001 - 0027, © European Communities, 1996; Figure 72: The Apple Museum; Figures 73(a): Lever Faberge, 73(b): Trademark and copyright of Cadbury Limited, reproduced with the permission of trademark owner, 73(c): Ferrari; Figure 74: Neill Bruce; Figure 77: General Motors; Figure 78: (top left): Photo by Steve Garner of drawing by Norman Edwards, a student of Loughborough University; (top right): Alfred Pasieka/Science Photo Library; (bottom right): Photo by Steve Garner, student designer unknown; (bottom left): Photo by Mark Evans, car design by Angus Davidson, a student of Loughborough University; Figure 83: Copyright Verner Panton Design, Switzerland; Figure 93: Photo: Mr Hans Hansen by courtesy of Vitra (www.vitra.com); Figure 113: Bayer plc; Figure 114: Christian Hicks; Figures 115, 116, 117, 118(a), (b) and 119: Penoyre & Prasad; Figure 121: Perkins Engines Co Ltd.; Figures 122(a) and (b): The Motoring Picture Library, The National Motor Museum; Figure 123: Ogle Noor; Figures 124(a): Robin Taylor; (b) Steve Behr/ Stockfile; (c) Nottinghamshire County Council: Community Services: Nottinghamshire Archives.

Every effort has been made to trace all copyright owners, but if any has been inadvertently overlooked, the publishers will be pleased to make the necessary arrangements at the first opportunity.

Course Team

Academic staff

Ken Baynes, External Assessor
Catherine Cooke, Author
Nigel Cross, Author
Chris Earl, Author
Steve Garner, Author and Course Chair
Georgy Holden, Author
Robin Roy, Author

Consultants

Mike Ashby, Contributing Author, Block 5
Mark Evans, Contributing Author, Workbook 1
Tony Hodgson, Contributing Author, Block 5

Associate lecturers

Jenny Burke
Nick Jeffrey

Course managers

Andy Harding, Course Manager
Amber Thomas, Course Manager

Production staff

Tammy Alexander, Graphic Designer
Margaret Barnes, Course Secretary
Deirdre Bethune, Course Secretary
Philippa Broadbent, Print Buyer
Jane Bromley, Interactive Media Designer
Michael Brown, Video Editor
Daphne Cross, Assistant Print Buyer
Tony Duggan, Learning Projects Manager
Bernie D'Souza, Course Secretary
Barbara Fraser, Picture Researcher
Phil Gauron, Video Producer
Richard Hearne, Photographer
Katie Meade, Rights Executive
Jane Moore, Editor
Jonathon Owen, Graphic Artist
Alex Reid, Narrator (video)
Ekkehard Thumm, Media Project Manager
Howie Twiner, Graphic Artist
Robert Wood, Editor